Principles of Plant Biotechnology

Principles of Plant Biotechnology

Jason Angstrom

R CALLISTO
REFERENCE

www.callistoreference.com

Callisto Reference,
118-35 Queens Blvd., Suite 400,
Forest Hills, NY 11375, USA

Visit us on the World Wide Web at:
www.callistoreference.com

ISBN: 978-1-64116-225-8 (Hardback)

Cataloging-in-Publication Data

Principles of plant biotechnology / Jason Angstrom.
    p. cm.
Includes bibliographical references and index.
ISBN 978-1-64116-225-8
1. Plant biotechnology. 2. Plants. 3. Agricultural biotechnology. I. Angstrom, Jason.
TP248.27.P55 P75 2019
631.523 3--dc23

Table of Contents

Permissions

Index

Preface

The use of living organisms to make or develop or modify products is under the broad field of biotechnology. Plant biotechnology is a branch of this discipline that is concerned with the application of the techniques of biotechnology for plant breeding and improvement. Some of the objectives include improving plant quality, increasing crop yield, increasing tolerance to environmental stresses, viruses, fungi, bacteria and pests. Such modifications are of immense use in agriculture. The techniques of marker assisted selection, doubled haploidy, reverse breeding and genetic modification facilitate such changes. This book is compiled in such a manner, that it will provide in-depth knowledge about the theory and practice of plant biotechnology. It aims to shed light on some of the unexplored aspects of this field. This book is an essential guide for both academicians and those who wish to pursue this discipline further.

Given below is the chapter wise description of the book:

Chapter 1- The application of the techniques of biotechnology for producing desired traits in plants is under the scope of plant biotechnology. Plants may be modified to improve their nutritional value for human consumption. This chapter has been carefully written to provide an introduction to plant biotechnology and discusses its fundamental concepts and principles.

Chapter 2- A plant cell is an eukaryotic cell with distinct characteristics such as the presence of a cell wall that has cellulose, hemicellulose and pectin, plastids, a large vacuole, etc. Plant tissues provide mechanical strength to plants. They are of two types, meristematic and permanent tissue. The topics elucidated in this chapter cover some of the important aspects of plant cell and tissue culture.

Chapter 3- Micropropagation is a vegetative propagation technique for the production of a number of pathogen-free and genetically-superior transplants in a limited space in a limited time. A detailed study of the fundamental aspects of micropropagation has been provided in this chapter, particularly meristem, callus, embryo, suspension and protoplast culture.

Chapter 4- Ploidy refers to the number of complete sets of chromosomes in a cell. This chapter explores ploidy in plants through a detailed discussion of ploidy, haploid plants, diploid plants, plant life cycles and polyploidy plants.

Chapter 5- Genetically modified crops are the plants that are used in agriculture for which the DNA has been modified using techniques of genetic engineering. A new trait is introduced in a plant or crop such as resistance to diseases, pests or environmental conditions, resistance to chemical treatments, reduction of spoilage, etc. This chapter discusses in detail the processes of gene gun method, protoplast isolation, microinjection, etc.

Chapter 6- Various microtechniques are used in plant biotechnology for the identification and characterization of plant species, which have significant implications in quality control and various other uses. This chapter carefully analyzes the varied microtechniques used in plant biotechnology, such as microscopy, cell sorting, plant histological technology, etc.

Chapter 7- A biofertilizer is a substance that consists of microorganisms, which is applied to plants to promote growth. The microbes act to colonize the rhizosphere and the plant's interior, and also increase the supply of primary nutrients to the plant. The aim of this chapter is to explore the use and production of biofertilizers, and its varied types such as phosphorus biofertilizers, compost biofertilizers and nitrogen biofertilizers.

At the end, I would like to thank all those who dedicated their time and efforts for the successful completion of this book. I also wish to convey my gratitude towards my friends and family who supported me at every step.

Jason Angstrom

Chapter 1

Introduction to Plant Biotechnology

The application of the techniques of biotechnology for producing desired traits in plants is under the scope of plant biotechnology. Plants may be modified to improve their nutritional value for human consumption. This chapter has been carefully written to provide an introduction to plant biotechnology and discuss its fundamental concepts and principles.

Plant biotechnology is a set of techniques used to adapt plants for specific needs or opportunities. Situations that combine multiple needs and opportunities are common. For example, a single crop may be required to provide sustainable food and healthful nutrition, protection of the environment, and opportunities for jobs and income. Finding or developing suitable plants is typically a highly complex challenge.

Plant biotechnologies that assist in developing new varieties and traits include genetics and genomics, marker-assisted selection (MAS), and transgenic (genetic engineered) crops. These biotechnologies allow researchers to detect and map genes, discover their functions, select for specific genes in genetic resources and breeding, and transfer genes for specific traits into plants where they are needed.

The roots of plant biotechnology can be traced back to the time when humans started collecting seeds from their favorite wild plants and began cultivating them in tended fields. It appears that when the plants were harvested, the seeds of the most desirable plants were retained and replanted the next growing season.

While these primitive agriculturists did not have extensive knowledge of the life sciences, they evidently did understand the basic principles of collecting and replanting the seeds of any naturally occurring variant plants with improved qualities, such as those with the largest fruits or the highest yield, in a process that we call artificial selection. This domestication and controlled improvement of plant species was the beginning of plant biotechnology.

This very simple process of selectively breeding naturally occurring variants with observably improved qualities served as the basis of agriculture for thousands of years and resulted in thousands of domesticated plant cultivars that no longer resembled the wild plants from which they descended.

The second era of plant biotechnology began in the late 1800's as the base of knowledge derived from the study of the life sciences increased dramatically.

In the 1860's Gregor Mendel, using data obtained from controlled pea breeding experiments, deduced some basic principles of genetics and presented these in a short monograph modestly titled "Versuche über Pflanzenhybriden".

Mendel proposed that heritable genetic factors segregate during sexual reproduction of plants and that factors for different traits assort independently of each other.

Mendel's work suggested a mechanism of heritable factors that could be manipulated by controlled breeding of plants through selective fertilization and also suggested that the pattern of inheritance for these factors could be analyzed or, in some cases, predicted by the use of mathematical statistics.

These findings complemented the work of Charles Darwin, who expounded the principles of descent with modification and selection as the chief factor of evolutionary change in his 1859 book "On the Origin of Species by Means of Natural Selection".

The application of these principles to agriculture resulted in deliberately produced hybrid varieties for a large number of cultivated plants via selective fertilization. These artificially selected hybrids soon began to benefit humankind with tremendous increases in both the productivity and the quality of food crops.

The third era of plant biotechnology involves a drastic change in the way crop improvement may be accomplished, by direct manipulation of genetic elements (genes).

This process is known as genetic engineering and results in plants that are called genetically modified organisms (GMOs), to distinguish them from plants that are produced by conventional plant-breeding methods.

Genetically modified plants can contribute desirable genes from outside traditional breeding boundaries. Even genes from outside the plant kingdom can now be brought into plants. For example, animal genes, including human genes, have been transferred into plants, a feat not replicated in nature.

Public Concern

It is perhaps this lack of natural boundaries for genetic exchange that seems so foreign to conventional scientific thought and that makes plant genetic engineering controversial.

The thought of taking genes from animals, bacteria, viruses, or any other organism and putting them into plants, especially plants consumed for food, has raised a host of questions among concerned scientists and public alike.

Negative public perception of genetically modified crops has affected the development and commercialization of many plant biotechnology products, especially food plants. While there are dozens of genetically engineered plants ready for field production, public pressure has delayed the release of some of these plants and has caused the withdrawal of others from the marketplace.

This public concern also appears to be driving increased government review of products and decreased government funding for plant biotechnology projects in Europe. Negative public perceptions do not seem to be as strong in Asia, since the pressures of feeding large populations tend to outweigh the perceived risks.

Economic Goals

To what end are humans genetically engineering plants? This is an essential question for researchers, executives of biotechnology companies, and consumers at large.

Before addressing technical questions about how to apply biotechnology, the desired goals must be clearly defined. The general goals of plant biotechnology appear to be:

- Economic improvement of existing products,

- Improvement of human nutrition, and

- Development of novel products from plants.

Economic improvements include increases in yield, quality, pest resistance, nutritional value, harvestability, or any other change that adds value to an established agricultural product.

Examples of this category include insect-protected tomatoes, potatoes, cotton, and corn; herbicide-resistant canola, corn, cotton, flax, and soybeans; canola and soybeans with genetically altered oil compositions; virus-resistant squash and papayas; and improved ripening tomatoes. All these examples were introduced to agriculture in the later half of the 1990's.

Nutritional Goals

Additionally, some products appearing in the scientific literature but awaiting commercialization have the potential to dramatically improve human nutritional deficiencies, which are especially prevalent in developing countries.

These products include "golden rice," genetically modified rice that produces carotenoids, a dietary source of vitamin A. Golden rice has the potential to prevent vitamin A deficiency in developing countries, where this vitamin deficiency is a leading cause of blindness.

Researchers are also using genetic engineering to increase the amount of the iron-storing protein ferritin in seed crops such as legumes. Iron deficiency, which affects 30 percent of the human population, can impair cognitive development and cause other other health problems. This proposed enhancement of iron content in consumable plant products could help more than a billion people who suffer from chronic iron deficiency.

Plant Tissue Cultures

Central to plant biotechnology is the use of in vitro methods. Researchers use plant tissue cultures, for example, to grow plant cells on sterile nutrient media. Countless recipes for these nutrient media exist. The choice of which one to use is based on the plant species and the tissue type to be grown.

All such media contain at least some of the important nutitional elements, such as nitrogen, potassium, calcium, magnesium, sulfur, phosphorus, iron, boron, manganese, zinc, iodine, molybdenum, copper, and cobalt, usually in the form of inorganic salts or as metal chelates, and an organic energy source, such as sucrose. The media may also contain vitamins, hormones, and other ingredients, depending on the intended use.

To initiate plant tissue culture, a piece of a living plant is excised and disinfected using a chemical disinfectant.

This piece of plant tissue, called an explant, is placed on a sterile plant tissue culture medium to grow. Many plant tissues may be used to obtain explants for plant tissue culture, including those from leaves, petioles, shoots, tubers, roots, and meristematic regions.

When an explant is placed in the sterile tissue culture medium, cells that are not terminally differentiated will grow and divide. If plant hormones are included in the recipe, the plant cells can be coaxed to develop into different types of tissues or organs.

By using a succession of media containing different hormones, it is possible to regenerate whole plants from single cells. The choice of tissue used for the explant and the choice of hormones included in the tissue culture medium depend on the desired result.

Micropropagation

Micropropagation, another biotechnology technique, is the production of many clonal plants using tissue culture methods. By means of micropropagation, it is possible to generate many thousands of plant clones using tissue explants obtained froma single parent plant. The main advantage to micropropagation is the potential of producing thousands of exact copies of a plant with desirable traits.

Micropropagation is especially important for rare plants, genetically engineered plants, and plants that have sexual reproductive problems. Many plant species are now routinely propagated by micro propagation methods, including orchids, ferns, many flowering ornamentals, and vegetable plants.

Steps in Genetic Engineering

The first genetically engineered plants, tobacco plants, were reported in the scientific literature in 1984. Since 1984 there have been thousands of genetically engineered plants produced in laboratories worldwide. The process of genetically engineering a plant involves several key steps:

- Isolating the genetic sequence (gene) to be placed from its biological source

- Placing the gene in an appropriate vehicle to facilitate insertion into plant cells

- Inserting the gene into the plant in a process known as plant transformation

- Selecting the few plant cells that contain the new gene (transformed cells) out of all the plant cells in the explant

- Multiplying the transformed cells in sterile tissue culture

- Regenerating the transformed cells into a whole plant that can grow outside the tissue culture vessel.

The gene or genes to be placed in the plant may be obtained from virtually any biological source: animals, bacteria, fungi, viruses, or other plants. Placing genes into an appropriate vehicle for transfer into a plant involves using various molecular biology techniques, such as restriction enzymes and ligation, to essentially "cut and paste" the gene or genes of interest into another DNA molecule, which serves as the transfer vehicle (vector).

Plant Transformation Methods

Currently plant transformation with foreign genes may be accomplished by several proven methods, including bacteria-mediated transfer, microparticle bombardment, electroporation, microinjection, sonication, and chemical treatment.

By far, the most often utilized method of plant transformation involves the use of naturally oc-curing plant pathogenic bacteria from the genus Agrobacterium. In nature, this bacterium infects plants and transfers some of its own bacterial DNA into the plant.

Through the action of proteins produced by the bacteria, bacterial DNA is made to integrate per-manently into the plant's own genomic DNA. Expression of the bacterial DNA in the plant causes the plant to produce unusual quantities of plant hormones and other compounds, called opines, which provide food for the bacteria.

The unusual quantities of plant hormones around the infection site cause the plant cells to grow abnormally, producing characteristic tumors. Scientists have harnessed this pathogenic bacterium to insert genes into plants by deleting the bacterial genes that cause tumors in the plant and then inserting desirable genes in their place.

When the modified Agrobacterium infects a plant, it transfers the desirable genes into the plant genome instead of causing tumors. The desirable genes become a permanent part of the plant ge-nome, and expression of these genes in plant cells produces desirable products.

One major drawback of the Agrobacterium method is that insertion of bacterial DNA into the plant genome is essentially random. The gene may not be efficiently transcribed at its location, or the insertion of bacterial DNA may knock out an important plant gene by inserting in the middle of it or both may occur. Therefore, the fact that a cell is genetically transformed does not guarantee that it will perform as desired.

Microparticle bombardment is the introduction of foreign DNA constructs into plant cells by at-taching the DNA to small metal particles and blasting the particles into plant cells using either a compressed air gun or a gun powered by a 0.22 caliber gun cartridge.

This is truly a "brute force" method of introducing DNA into a cell that inadvertently causes many lethal casualties among the bombarded plant cells. However, some plant cells blasted with the DNA-containing metal particles will recover and survive.

The plant cells may express the DNA for only a short time (transient expression), because the DNA does not readily integrate into the plant genome, but occasionally the foreign DNA may sponta-neously recombine into the plant genome and become permanent.

Other ways of introducing foreign DNA into plant cells include electroporation, microinjection, sonication, and chemical treatment. These methods are not used extensively, because they gen-erally require the production of protoplasts (plant cells that lack their cell walls) from plant cells before transformation.

To create protoplasts, the plant cell wall is removed by digestion with the enzymes cellulase and pectinase. Protoplasts are fragile structures, but the absence of a cell wall is desirable because it leaves only the plasma membrane as a barrier to foreign DNA entering a plant cell.

Electroporation uses very brief pulses of high-voltage electrical energy to create temporary holes in the plasma membrane through which the foreign DNA can pass. Micro injection in-volves physically injecting a small amount of DNA into a plant cell using a microscope and an extremely fine needle.

Sonication uses ultrasonic waves to punch temporary holes in the plasma membrane; this method is therefore similar to electroporation. Chemical treatment involves the use of polyethylene glycol to render the plasma membrane permeable to foreign DNA.

All the transformation procedures produce only a few transformed cells out of the millions of cells in an explant, so selection of transformed cells is essential.

Selection of Transformed Plant Cells

Selecting the few transformed plant cells out of all the plant cells in an explant requires some advance planning. Most foreign DNA constructs introduced into a plant are designed and built to contain additional genes that function as selectable markers or reporter genes. Selectable markers include genes for resistance to antibiotics or herbicides.

Plant cells containing and expressing these genes will be tolerant of antibiotics or herbicides added to the plant tissue culture media, while the non transformed plant cells will be killed off. The surviving cells in the tissue culture media are mostly transformed.

Instead of selectable markers, reporter genes may be used. Reporter genes induce an easily observable trait to transformed plant cells that facilitates the physical isolation of these cells.

Reporter genes include beta-glucuronidase, luciferase, and plant pigment genes. Beta-glucuronidase (commonly known as GUS) allows the plant cells expressing this gene to metabolize colorigenic substrates while non transformed plant cells cannot.

To use this test, researchers treat a small amount of plant tissue with the colorigenic chemical substrate. If the cell turns color (blue) it is known to be transformed and expressing the GUS gene. If the cell does not turn color, it probably is not transformed.

Another reporter gene is luciferase, an enzyme isolated from fireflies. Luciferase makes plant cells glow in the presence of certain chemicals if the gene is present; hence, transformed cells glow, where as non transformed cells do not glow.

Plant pigment genes, such as anthocyanin pigment genes, occur naturally in plants and produce pigments that impart color to flowers. Inclusion of these pigment genes as reporter genes will allow transformed plant cells to be selected by their color.

Transformed cells have color, while non transformed cells remain colorless. Both selectable markers and reporter genes allow selection of cells into which genes have been successfully inserted and are operating properly.

Regenerating Whole Transformed Plants

After successfully getting a gene construct into a plant cell and selecting the transformed cells, it is possible to get the plant cells to multiply in tissue culture. Also, by treating the plant cells with combinations of plant hormones, the cells are made to differentiate into various plant organs or whole plants.

For example, treating transformed plant cells with a high concentration of the plant hormone cytokinin causes shoots to develop. Transferring these shoots to anothermedium, one that is high in the plant hormone auxin, will cause roots to develop on the shoots.

In this way a whole transgenic plant may be regenerated from transformed plant cells. Once a transformed plant is regenerated in tissue culture, the plant may be transferred to a climate-controlled greenhouse, where it can grow to maturity.

Future generations of transgenic plants may then be propagated sexually via seeds or asexually via vegetative propagation methods. Often transgenic plants must be grown in containment greenhouses to prevent accidental release into the environment.

In such high-tech greenhouses, all factors contributing to optimal plant growth—lighting, temperature, humidity, nutrients, and other environmental conditions—are tightly controlled. Often hydroponic systems, which use a solution of plant nutrients as a growth medium in place of soil, are employed to control all aspects of plant nutrition.

Chapter 2

Plant Cell and Tissue Culture

A plant cell is a eukaryotic cell with distinct characteristics such as the presence of a cell wall that has cellulose, hemicellulose and pectin, plastids, a large vacuole, etc. Plant tissues provide mechanical strength to plants. They are of two types, meristematic and permanent tissue. The topics elucidated in this chapter cover some of the important aspects of plant cell and tissue culture.

Plant cells are the basic unit of life in organisms of the kingdom Plantae. They are eukaryotic cells, which have a true nucleus along with specialized structures called organelles that carry out different functions. Plant cells are differentiated from the cells of other organisms by their cell walls, chloroplasts, and central vacuole.

Plant Cell Structure

The plant cell has many different parts. Each part of the cell has a specialized function. These structures are called organelles.

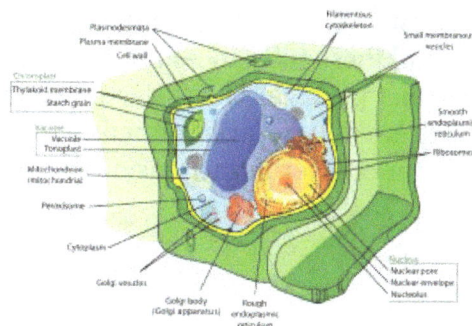

This figure shows the various parts of a plant cell. Specialized structures in plant cells include chloroplasts, a large vacuole, and the cell wall.

Chloroplasts

Chloroplasts are found only in plant and algae cells. These organelles carry out the process of photosynthesis, which turns water, carbon dioxide, and light energy into nutrients. They are oval-shaped and have two membranes: an outer membrane, which forms the external surface of the chloroplast, and an inner membrane that lies just beneath. Between the outer and inner membrane is a thin intermembrane space about 10-20 nanometers wide. Within the other membrane, there is another space called the stroma, which is where chloroplasts are contained. Chloroplasts themselves contain many flattened disks called thylakoids, and these have a high concentration of chlorophyll and carotenoids, which capture light energy. The molecule chlorophyll also gives plants their green color. Thylakoids are stacked on top of one another in vascular plants in stacks called grana.

Vacuoles

Plant cells are unique in that they have a large central vacuole. A vacuole is a small sphere of membrane within the cell that can contain fluid, ions, and other molecules. Vacuoles are basically large vesicles. They can be found in the cells of many different organisms, but plant cells characteristically have a large vacuole that can take up anywhere from 30-80 percent of the cell.

The central vacuole of a plant cell helps maintain its turgor pressure, which is the pressure of the contents of the cell pushing against the cell wall. A plant thrives best when its cells have high turgidity, and this occurs when the central vacuole is full of water. If turgor pressure in the plants decreases, the plants begin to wilt. Plant cells fare best in hypotonic solutions, where there is more water in the environment than in the cell; under these conditions, water rushes into the cell by osmosis, and turgidity is high. Animal cells, on the other hand, can lyse if too much water rushes in; they fare better in isotonic solutions, where the concentration of solutes in the cell and in the environment is equal and net movement of water in and out of the cell is the same.

Cell Wall

The cell wall is a tough layer found on the outside of the plant cell that gives it strength and also maintains high turgidity. In plants, the cell wall contains mainly cellulose, along with other molecules like hemicellulose, pectin, and liginins. The composition of the plant cell wall differentiates it from the cell walls of other organisms. For example, fungi cell walls contain chitin, and bacterial cell walls contain peptidoglycan, and these substances are not found in plants. A main difference between plant and animal cells is that plant cells have a cell wall while animal cells do not. Plant cells have a primary cell wall, which is a flexible layer formed on the outside of a growing plant cell, and a secondary cell wall, a tough, thick layer formed inside the primary plant cell wall when the cell is mature.

Cell Membrane

This is a thin and a semi-permeable membrane that encloses the cell's contents. This thin lining will be present at the inner side of the plant cell wall.

Cytoplasm

This is a gel-like fluid that is found within the cell membrane. This will contain various elements like:

- Water
- Enzymes
- Salts
- Organelles
- Organic molecules.

Cytoskeleton

This is basically a bundle of fibers that are found throughout the cytoplasm. The role of the cytoskeleton is to help the cell maintain the shape. It also provides support to the cell.

Endoplasmic Reticulum

This is a wide network of membranes. There are two types of regions in endoplasmic reticulum – one with ribosomes and other without ribosomes. The former is called rough endoplasmic reticulum and the latter is called smooth endoplasmic. The main function of this organelle is protein and lipid synthesis.

Golgi Complex

This apparatus, which is found in all eukaryotic cells, is involved in distributing macromolecules to different parts of the cell.

Microtubules

These are the hollow rods whose primary function is to provide support, as well as shape to the cell. These have important roles to play in the chromosome movement during plant cell division.

Mitochondria

This is the "powerhouse" of the cell meaning that this generates energy for the cell through the process called respiration.

Nucleus

As mentioned earlier, this is the membrane-bound organelle and contains the cell's DNA. The nucleus contains nucleolus that helps in the production of ribosomes and nucleopore that allows the transport of nucleic acids and proteins from and to the nucleus.

Peroxisomes

These are small structures, which are enveloped by a single membrane that contains the enzyme. The peroxisomes will be involved in plant processes like photorespiration.

Plasmodesmata

These are the channels that are located between cell walls, which allow for the molecules and signals to pass between the plant cells.

Ribosomes

These are the organelles that contain RNA and protein elements. As such, the ribosomes will be responsible for the protein assembly. Within the plant cell, the ribosomes are either found to be adhering to the endoplasmic reticulum or floating in the cytoplasm.

Three different types of Plant Cell

Cells of a matured plant become specialized to perform certain vital functions that are essential for survival. Some plants cells are used for transferring nutrients and for storing food. Beneath are some of the specialized plant cells.

- Collenchyma cells: It acts as a supporting system when there is a restraining growth in a plant due to lack of hardening agent in primary walls.

- Sclerenchyma cells: These cells are more rigid compared to collenchyma cells. Sclerenchyma cells consist of a hardening agent and their main function is to provide support to the plants.

- Parenchyma cells: These cells are used to store organic products in plants.

Tissue

Plants are made up of tissue, just like any other living organism. Also like other living organisms, plants have several different types of tissues. In humans, your skin is a different type of tissue than muscles, and organ tissue is different from ligament tissue. Plants follow the same rule.

Plant tissue systems fall into one of two general types: meristematic tissue and permanent (or non-meristematic) tissue. Cells of the meristematic tissue are found in meristems, which are plant regions of continuous cell division and growth. Meristematic tissue cells are either undifferentiated or incompletely differentiated, and they continue to divide and contribute to the growth of the plant. In contrast, permanent tissue consists of plant cells that are no longer actively dividing.

Meristematic tissues consist of three types, based on their location in the plant. Apical meristems contain meristematic tissue located at the tips of stems and roots, which enable a plant to extend in length. Lateral meristems facilitate growth in thickness or girth in a maturing plant. Intercalary meristems occur only in monocots, at the bases of leaf blades and at nodes (the areas where leaves attach to a stem). This tissue enables the monocot leaf blade to increase in length from the leaf base; for example, it allows lawn grass leaves to elongate even after repeated mowing.

Meristems produce cells that quickly differentiate, or specialize, and become permanent tissue. Such cells take on specific roles and lose their ability to divide further. They differentiate into three main types: dermal, vascular, and ground tissue. Dermal tissue covers and protects the plant, and vascular tissue transports water, minerals, and sugars to different parts of the plant. Ground tissue serves as a site for photosynthesis, provides a supporting matrix for the vascular tissue, and helps to store water and sugars.

Secondary tissues are either simple (composed of similar cell types) or complex (composed of different cell types). Dermal tissue, for example, is a simple tissue that covers the outer surface of the plant and controls gas exchange. Vascular tissue is an example of a complex tissue, and is made of two specialized conducting tissues: xylem and phloem. Xylem tissue transports water and nutrients from the roots to different parts of the plant, and includes three different cell types: vessel elements and tracheids (both of which conduct water), and xylem parenchyma. Phloem tissue, which transports organic compounds from the site of photosynthesis to other parts of the plant, consists of four different cell types: sieve cells (which conduct photosynthates), companion cells, phloem parenchyma, and phloem fibers. Unlike xylem conducting cells, phloem conducting cells are alive at maturity. The xylem and phloem always lie adjacent to each other. In stems, the xylem and the phloem form a structure called a vascular bundle; in roots, this is termed the vascular stele or vascular cylinder.

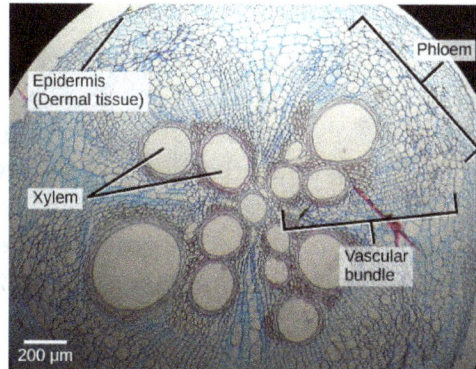

Figure. This light micrograph shows a cross section of a squash (Curcurbita maxima) stem.

Each teardrop-shaped vascular bundle consists of large xylem vessels toward the inside and smaller phloem cells toward the outside. Xylem cells, which transport water and nutrients from the roots to the rest of the plant, are dead at functional maturity. Phloem cells, which transport sugars and other organic compounds from photosynthetic tissue to the rest of the plant, are living. The vascular bundles are encased in ground tissue and surrounded by dermal tissue

All three types of plant cells are found in most plant tissues. Three major types of plant tissues are dermal, ground, and vascular tissues.

Dermal Tissue

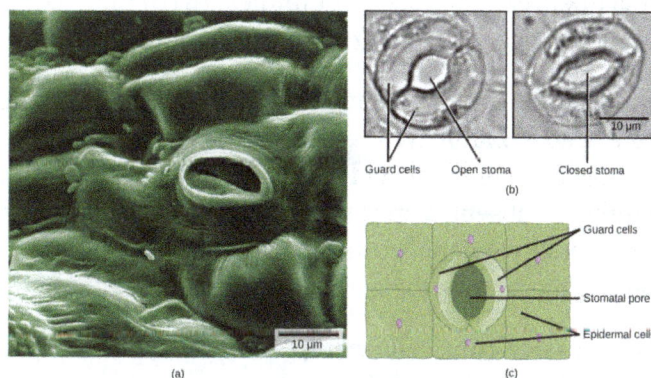

Figure. Openings called stomata (singular: stoma) allow a plant to take up carbon dioxide and release oxygen and water vapor. The (a) colorized scanning-electron micrograph shows a closed stoma of a dicot. Each stoma is flanked by two guard cells that regulate its (b) opening and closing. The (c) guard cells sit within the layer of epidermal cells

The dermal tissue of the stem consists primarily of epidermis, a single layer of cells covering and protecting the underlying tissue. Woody plants have a tough, waterproof outer layer of cork cells commonly known as bark, which further protects the plant from damage. Epidermal cells are the most numerous and least differentiated of the cells in the epidermis. The epidermis of a leaf also contains openings known as stomata, through which the exchange of gases takes place. Two cells, known as guard cells, surround each leaf stoma, controlling its opening and closing and thus regulating the uptake of carbon dioxide and the release of oxygen and water vapor. Trichomes are hair-like structures on the epidermal surface. They help to reduce transpiration (the loss of water by aboveground plant parts), increase solar reflectance, and store compounds that defend the leaves against predation by herbivores.

Vascular Tissue

The xylem and phloem that make up the vascular tissue of the stem are arranged in distinct strands called vascular bundles, which run up and down the length of the stem. When the stem is viewed in cross section, the vascular bundles of dicot stems are arranged in a ring. In plants with stems that live for more than one year, the individual bundles grow together and produce the characteristic growth rings. In monocot stems, the vascular bundles are randomly scattered throughout the ground tissue.

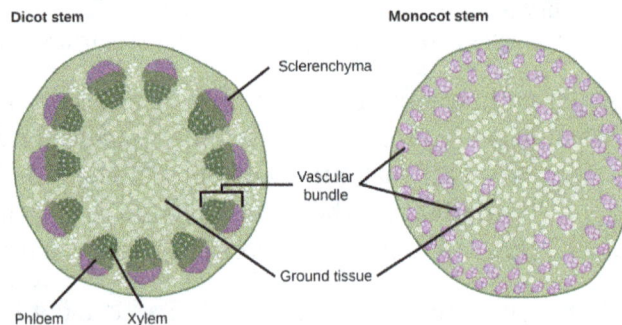

In (a) dicot stems, vascular bundles are arranged around the periphery of the ground tissue. The xylem tissue is located toward the interior of the vascular bundle, and phloem is located toward the exterior. Sclerenchyma fibers cap the vascular bundles. In (b) monocot stems, vascular bundles composed of xylem and phloem tissues are scattered throughout the ground tissue

Xylem tissue has three types of cells: xylem parenchyma, tracheids, and vessel elements. The latter two types conduct water and are dead at maturity. Tracheids are xylem cells with thick secondary cell walls that are lignified. Water moves from one tracheid to another through regions on the side walls known as pits, where secondary walls are absent. Vessel elements are xylem cells with thinner walls; they are shorter than tracheids. Each vessel element is connected to the next by means of a perforation plate at the end walls of the element. Water moves through the perforation plates to travel up the plant.

Phloem tissue is composed of sieve-tube cells, companion cells, phloem parenchyma, and phloem fibers. A series of sieve-tube cells (also called sieve-tube elements) are arranged end to end to make up a long sieve tube, which transports organic substances such as sugars and amino acids. The sugars flow from one sieve-tube cell to the next through perforated sieve plates, which are found at the end junctions between two cells. Although still alive at maturity, the nucleus and other cell components of the sieve-tube cells have disintegrated. Companion cells are found alongside the sieve-tube cells, providing them with metabolic support. The companion cells contain more ribosomes and mitochondria than the sieve-tube cells, which lack some cellular organelles.

Ground Tissue

Ground tissue is mostly made up of parenchyma cells, but may also contain collenchyma and sclerenchyma cells that help support the stem. The ground tissue towards the interior of the vascular tissue in a stem or root is known as pith, while the layer of tissue between the vascular tissue and the epidermis is known as the cortex.

Plant Cell Culture

Plant cell cultures are living plant cells grown in a lab. They originate from living plants, typically grown in natural conditions. Scientists take scrapings of a living mother plant, locate the specific cells that they want to duplicate, and create the perfect environment for those types of cells to grow in their culture. For the cells to grow effectively, the cultures need the exact vitamins, nutrients, and hormone content that exists in the growing plant form.

The exact living circumstances are carefully recreated in an aqueous solution, then the plant cells are taken from the mother plant and placed in the solution to multiply. Once they have multiplied enough, they are removed from the solution and suspended in their active state.

This budding science can be used to grow plants in a lab that can eventually be planted in soil and cared for as any other plant would be.

The precise composition of the culture medium will depend on what is required. Cell proliferation has different requirements to cell differentiation. However, basic growth is supported by a basic medium, which is generally composed of water, macro- and micro-nutrients, and a carbohydrate source, usually sucrose, to replace the carbon, which the plant normally fixes from the atmosphere by photosynthesis. To improve growth, media can include trace amounts of certain organic compounds, such as vitamins, amino acids and plant growth regulators. Culture media can also contain what are known as undefined components. These include fruit juices, plant extracts, yeast extract etc. One of the most well-known of these undefined components is coconut milk, which has been a popular addition for orchid culture. Although, good results can be achieved using these ingredients, their use is not encouraged as their composition is not consistent, and can vary each time they are used. Depending on whether or not the culture medium is to be solid or liquid, a gelling agent is added.

95 percent of a culture medium is water, and therefore the quality of the water is important. It is recommended that the water is always distilled and in some cases, double-distilled. Macro- and micro-nutrients are essential for growth, and so are found in all culture media.

When the nutrient requirements of a plant are unknown, the nutrient composition as defined by Murashige and Skoog, can be used, providing that the plant is not sensitive to salt. Sugar is a very important component in any culture medium. A concentration of 1-5 percent saccharose (a disaccharide) is usually used, as this sugar is synthesized and transported naturally by the plant. The gelling agent usually used in culture media is agar. Agar, a seaweed derivative, is a polysaccharide with a high molecular mass. For medium of the optimum solidity, agar is usually added at a concentration between 0.6-0.8 percent.

Growth regulators are added according to what is required from the culture, but also depends on the type of explant and the plant species. For example, eggplants which themselves produce sufficient auxin do not need extra auxin for cell extension and/or division. Similarly, with explants producing enough cytokinin, cytokinin will not have to be added to the medium.

There are general rules that apply with the use of growth regulators but the individuality of the explant will always have some influence. Generally auxins cause cell elongation and expansion, cell division and the formation of adventitious roots, inhibition of adventitious and axillary shoot formation, and embryogenesis in suspension cultures. With low auxin concentration, adventitious root formation occurs whereas if the auxin concentration is high, callus formation is a possibility.

Cytokinins stimulate growth and development. In high concentrations they can be used to induce adventitious shoot formation, but root formation is generally inhibited. Care should be taken in the use of growth regulators as there is evidence in the literature that excessive use of certain growth regulators can lead to a large number of mutations.

Once the medium has been prepared it has to be at the correct pH. The pH used is between 5.5 to 6.0. If the pH is too high, this can stop growth and development. With a pH that is too low, gelling can be affected, as well as the uptake of some of the components. Finally the medium has to be sterilized, and this usually takes place in an autoclave. Providing exposure is sufficient, pressurized steam can destroy all microorganisms. The conditions for sterilization are 20 mins at 121°C and 15psi.

Plant Tissue Culture

Plant tissue culture is the technique of maintaining and growing plant cells, tissues or organs especially on artificial medium in suitable containers under controlled environmental conditions.

The part which is cultured is called explant, i.e., any part of a plant taken out and grown in a test tube, under sterile conditions in special nutrient media. This capacity to generate a whole plant from any cell/explant is called cellular toti-potency. In fact, the whole plant can be regenerated from any plant part (referred to as explant) or cells. Gottlieb Haberlandt first initiated tissue culture technique in 1902.

Hormones used in plant tissue culture are as follows:

1. Auxins neoline (Indole-3-acetic acid, Indole-3-butyric acid, Potassium Salt— Naphthalene acetic acid 2, 4-Dichlorophenoxyacetic acid p-Chloro-phenoxy acetic acid)

2. Cytokinins (6-Benzylaminopurine, 6-Dimethylallylaminopurine (2ip), Kinetin)

3. Gibberellins (Gibberellic Acid)

4. Abscisic Acid (ABA) (Abscisic Acid)

5. Polyamines (Putrescine, Spermidine).

Environmental Conditions

There are three important aspects in vitro (outside the living organism and in an artificial environment) culture namely:

1. Nutrient medium,

2. Aseptic conditions and

3. Aeration of the tissue.

Nutrient Medium

The composition of plant tissue culture medium can vary depending upon the type of plant tissues or cell that are used for culture. A typical (generalized) nutrient consists of inorganic salts (both micro and macro elements), a carbon source (usually sucrose), vitamins (e.g., nicotonic acid, thiamine, pyridoxine and myoinositol), amino acids (e.g., arginine) and growth regulators (e.g., auxins like 2,4-D or 2,4-dichlorophenoxyacetic acid and cytokinins such as BAP = benzlaminopurine and gibberellins). Other compounds like casein hydrolysate, coconut milk, malt extract, yeast extract, tomato juice, etc. may be added for specific purposes.

Plant hormones play important role in growth and differentiation of cultured cells and tissues. An optimum pH (usually 5.7) is also very important. The most extensively used nutrient medium is MS medium which was developed by Murashige and Skoog in 1962. Usually a gelling agent agar (a polysaccharide obtained from a red algae Gelidium amansi) is added to the liquid medium for its solidification.

Aseptic Conditions (Sterilization)

Nutrient medium contains ample sugar which increases growth of microorganisms such as bacteria and fungi. These microbes compete with growing tissue and finally kill it. It is essential to maintain aseptic conditions of tissue culture. Thus sterilization means complete destruction or killing of microorganisms so that complete aseptic conditions are created for in vitro culturing.

Aeration of the Tissue

Proper aeration of the cultured tissue is also an important aspect of culture technique. It is achieved by occasionally stirring the medium by sterring or by automatic shaker.

Plant material—the explant: Any part of a plant taken out and grown in test tube under sterile conditions in special nutrient media is called explant.

Methods of Plant Tissue Culture

Plant tissue culture includes two major methods:

1. Type of in vitro growth-callus and suspension cultures.

2. Type of explant— single cell culture, shoot and root cultures, somatic embryo culture, meristem culture, anther culture and haploid production, protoplast culture and somatic hybridisation, embryo culture, ovule culture, ovary culture, etc.

Types of Plant Tissue Culture

Callus and Suspension Cultures

In callus culture, cell division in explant forms a callus. Callus is irregular unorganised and undifferentiated mass of actively dividing cells. Darkness and solid medium gelled by agar stimulates callus formation. The medium ordinarily contains the auxin, 2,4-D, (2, 4- dichlorophenoxy acetic acid) and often a cytokinin like BAP (Benzyl aminopurine). Both are growth regulators. This stimulates cell divison in explant. Callus is obtained within 2-3 weeks.

A suspension culture consists of single cells and small groups of cells suspended in a liquid medium. Usually, the medium contains the auxin 2,4-D. Suspension cultures must be constantly agitated at 100-250 rpm (revolutions per minute). Suspension cultures grow much faster than callus culture.

Sub Culturing

If tissue cultures are kept in the same culture vessel, they die in due course of time. Therefore, cells/tissues are regularly transferred into new culture vessels containing fresh media. This process is called sub culturing. It is important to note that during subculture; only a part of the culture from a vessel is transferred into the new culture vessel.

The callus and suspension cultures may be used to achieve cell biomass production, regeneration of plantlets, production of transgenic plants and isolation of protoplasts.

Single Cell Culture (Cell Cloning)

Cells derived from a single cell through mitosis constitute a clone and the process of obtaining clones is called cloning (asexual progeny of a single individual make up a clone). There are two popular techniques for single cell culture.

1. Bergmann's Plating Technique:

    This is widely used technique. The cells are suspended in a liquid medium at a cell density that is twice the desired density in the plate. Sterilized agar (Ca 1%) medium is kept malted in a water bath at 35°C. Equal volumes of the liquid and agar media are mixed and spread in Ca 1 mm thick layer in a petridish. The cells remain embedded in the soft agar medium which is observable under a microscope. When large colonies develop they are isolated and cultured separately.

2. Filter Paper Raft Nurse Tissue Technique:

    Single cells are placed on small pieces (8×8 mm) of filter paper, which are placed on top of callus cultures several days in advance. This allows the filter papers to be wetted by the callus tissues. The single cells placed on the filter paper derive their nutrition from the callus. The cells divide and form macroscopic colonies on the filters. The colonies are isolated and cultured.

Shoot and Root Cultures

Shoot culture is promoted by a cytokinin like BAR However; root culture is promoted by an auxin like NAA (naphthalene acetic acid). The shoot and root cultures are generally controlled by auxin-cytokinin balance. Usually, an excess of auxin promotes root culture, whereas that of cytokinin promotes shoot culture. Roots culture from the lower end of these shoots to give complete plantlets.

Somatic Embryo Culture

A somatic embryo develops from a somatic cell. The pattern of development of a somatic embryo is comparable to that of a zygotic embryo. Somatic embryo culture is induced by a high concentration of an auxin, such as 2,4-D. These embryos develop into mature embryos. Mature somatic embryos or embryoids germinate to give complete plantlets.

Establishment in the Field

The plantlets are removed from culture vessels and established in the field. This transfer is done by specific procedures called hardening. During hardening, plantlets are kept under reduced light and high humidity. Hardening procedures make the plantlets capable of tolerating the relatively harsher environments outside the culture vessels.

Endosperm Culture

Tissue culture methods are also used for culturing endosperm. It is unique because it supplies nutrition to the developing embryo. It is also triploid in its chromosome constitution. Triploid plants are used for the production of seedless fruits (e.g., apple, banana etc.). The technique of endosperm culture involves the following:

1. The immature seeds are dissected under aseptic condition. Endosperms along with embryos, are excised. Sometimes, mature seeds can also be used.

2. The excised endosperms are cultured on a suitable medium and embryos are removed after initial growth.

3. The initial callus phase is developed.

4. The shoots and roots may develop and complete triploid plants are formed for further use.

Meristem Culture

Meristem is a localized group of cells, which are actively dividing and undifferentiated but ultimately giving rise to permanent tissue. Although the plant is infected with a virus, yet the meristem is free of virus. Therefore, meristem can be removed and grown in vitro to obtain virus free plants. Cultivation of axillary or apical shoot meristems is called meristem culture. The apical or axillary meristems are generally free from virus. Meristem culture involves the development of an already existing shoot meristem and subsequently, the regeneration of adventitious roots from the developed shoots.

It usually does not involve the regeneration of a new shoot meristem. The explants commonly used in meristem culture are shoot tips and nodal segments. These explants are cultured on a medium containing a cytokinin (generally BAP). The plantlets thus obtained are subjected to hardening and, ultimately, established in the fiddi Meristem culture is carried out in Potato, Banana, Cardamom, Orchids (protocorm stage), Sugarcane, Strawberry, Sweet Potato, etc. It is used in:

1. Production of virus-free plants like potato, sugarcane, banana and apple.

2. Germplasm conservation.

3. Production of transgenic plants.

4. Rapid clonal multiplication.

Anther Culture and Haploid Production

An individual/cell having the chromosome number found in the gametes of the species is called haploid. Formation of haploid is called haploid production. Thus haploid individuals arise from the gametes. A haploid has only one copy of each chromosome. Haploids are sterile and of no direct value.

When the chromosome number of a haploid plant is doubled, the plants of normal chromosome number for particular species are obtained. These plants are homozygous and are produced in 2-3 years. The chromosome number of these haploid plants is doubled by using colchicine to obtain homozygous plants.

In nature, haploid plants originate from unfertilized egg cells, but in laboratory, they can be produced from both male and female gametes. Anther is the part of the flower of Angiosperms producing pollen (microspores), borne at the end of the stamens and usually consisting of four sporangia. When anthers of some plants are cultured on a suitable medium to produce haploid plants, it is called anther culture.

The technique was developed by Guha and Maheshwari (1964) who cultured mature anthers of Datura innoxia. It is highly useful for the improvement of many crop plants. It is also useful for immediate expression of mutations and quick formation of purelines. This technique was first used in India to produce haploids of Datura. In many plants, haploids are also produced by culturing unfertilized ovaries/ovules. Sometimes, pollen grains are separated from anthers and cultured on suitable medium.

Embryo Culture

Culturing young embryos on a nutrient medium is called embryo culture. Young embryos are obtained from the developing seeds. The embryos complete their development on the medium and grow into seedlings. In general, older embryos are more easily cultured in vitro than young embryos.

Embryo culture is useful as follows:

1. Orchid seeds do not have any form of stored food. Embryos of such seeds can be cultured to obtain seedlings and maximum seedling formation can be achieved. Embryo culture in orchids can be applied for rapid clonal propagation.

2. In certain species, inhibitors present in the endosperm or seed coat make the seed dormant. Such embryos can escape dormancy by culturing on a suitable medium.

3. In certain hybrid seeds developed after interspecific crosses, the endosperm degenerates at an early stage and the young embryo is left with no food, consequently it also dies. Such young embryos can be excised from the seeds and cultured on the nutritive medium. Getting nutrition, they develop into seedlings which can be transplanted in the field.

4. A popular example includes hybridization of barley and wheat with Hordeum bulbosum leading to the production of haploid barley and haploid wheat respectively. Haploid wheat plants have also been successfully obtained through culture of hybrid embryos from wheat x maize crosses.

Ovule Culture

Ovule culture technique is utilized for raising hybrids which normally fail to develop due to the abortion of the embryos at an early stage. Ovules can easily be excised from the ovary and cultured on the basal medium. The loss of a hybrid embryo due to premature abscission of fruits may be prevented by ovule culture. In some cases, addition of fruit/vegetable juice increase the initial growth.

Ovary Culture

Ovary culture technique has also been successfully employed to raise interspecific hybrids between sexually incompatible species, Brassica campestris and B. oleracea. Ovaries are excised from the flowers and cultured at the zygote or two-celled proembryo stage for obtaining normal development on culture medium.

Sometimes coconut milk when used as a supplement to the medium promote formation of fruits that are larger than those formed in vivo (within the living organism). In Anethum, addition of kinetin in the medium caused polyembryony which gave rise to multiple shoots.

Micro Propagation

Micropropagation is the tissue culture technique used for rapid vegetative multiplication of ornamental plants and fruit trees by using small sized explants. Because of minute size of the propagules in the culture, the propagation technique is named as mircopropagation. This method of tissue culture produces several plants. Each of these plants will be genetically identical to the original plant from where they were grown.

The genetically identical plants developed from any part of a plant by tissue culture/micropropagation are called somaclones. The members of a single somaclone have the same genotype. This micropropagation is also known as somaclonal propagation. It is the only process adopted by Indian plant biotechnologists in different industries mainly for the commercial production of ornamental plants like lily, orchids, Euca lyptus, Cinchona, Blueberry, etc. and fruit trees like tomato, apple, banana, grapes, potato, citrus oil palm, etc.

There are four defined steps in micro propagation method. These are:

1. Initiation of culture from an explant like shoot tip on a suitable nutrient medium.

2. Shoot formation multiple shoots formation from the cultured explant.

3. Rooting of shoots rooting of in vitro developed shoots.

4. Transplantation the hardening of tissue culture raised plants and subsequent transplantation to the field.

Advantages of Micro Propagation

These are as follows:

1. It helps in rapid multiplication of plants.

2. A large number of plantlets are obtained within a short period and from a small space.

3. Plants are obtained throughout the year under controlled conditions, independent of seasons.

4. Sterile plants or plants which cannot maintain their characters by sexual reproduction are multiplied by this method.

5. It is an easy, safe and economical method for plant propagation.

6. In case of ornamentals, tissue culture plants give better growth, more flowers and less fall-out.

7. Genetically similar plants (somaclones) are formed by this method. Therefore, desirable characters (genetope) and desired sex of superior variety are kept constant for many generations.

8. The rare plant and endangered species are multiplied by this method and such plants are saved.

Regeneration of Plantlets

1. Preparation of suitable nutrient medium:

   Suitable nutrient medium as per objective of culture is prepared and transferred into suitable containers.

2. Selection of explants:

   Selection of explants such as shoot tip should be done.

3. Sterilisation of explants:

   Surface sterilization of the explants by disinfectants and then washing the explants with sterile distilled water is essential.

4. Inoculation:

   Inoculation (transfer) of the explants into the suitable nutrient medium (which is sterilized by filter-sterilized to avoid microbial contamination) in culture vessels under sterile conditions is done.

5. Incubation:

   Growing the culture in the growth chamber or plant tissue culture room, having the appropriate physical condition (i.e., artificial light; 16 hours of photoperiod), temperature (-26°C) and relative humidity (50-60%) is required.

6. Regeneration:

   Regeneration of plants from cultured plant tissues is carried out.

7. Hardening:

   Hardening is gradual exposure of plantlets to an environmental conditions.

8. Plantlet transfer:

   After hardening plantlets transferred to the green house or field conditions following acclimatization (hardening) of regenerated plants.

Protoplast Culture and Somatic Hybridisation

When a hybrid is produced by fusion of somatic cells of two varieties or species, it is known as somatic hybrid. The process of producing somatic hybrids is called somatic hybridisation. First, the cell wall of the plant cells is removed by digestion with a combination of pectinase and cellulase. The plant cells without cell wall are called protoplasts.

The protoplasts of the two plants are brought together and made to fuse in a solution of polyethylene glycol (PEG) or sodium nitrate. The fusion of protoplasts with the help of chemicals is called chemo-fusion. Fusion of protoplasts with the help of high voltage pulse is known as electro-fusion. The fusion of protoplasts not only involves the fusion of their cytoplasm but also their nuclei. The fused protoplasts are allowed to grow on culture medium. Soon they develop their own walls when they are called somatic hybrid cells.

The hybrid cells give rise to callus. Callus later differentiates into new plant which is somatic hybrid between two plants. Somatic hybrids in plants were first obtained between two species of Tobacco (Nicotiana glauca and N. langsdorfit) by Carlson et al in 1972. Successful somatic hybrids have also been got from different species of Brassica, Petunia, and Solanum.

Pomato is somatic hybrid between Potato and Tomato that belong to two different genera and Bomato is somatic hybrid between Brinjal and Tomato. Somatic hybrids are also produced between rice and carrot. The hybrid plant bears both fruits and tubers of the two parents.

- Protoplast technology has opened up avenues for development of hybrids of even asexually reproducing plants.

- There is a distinct possibility of development of new crop plants, e.g., Pomato.

- Somatic hybrids may be used for the production of useful allopolyploids (Individuals produced by interspecific polyploidy).

- Genetic manipulations can be carried out more rapidly when plant cells are in protoplast state. New genes can be introduced (e.g., male sterility, herbicide resistance). Mutations will be easier.

Artificial Seeds

There are many plants which neither have seeds nor produce a small quantity of seeds. To overcome this problem the concept of artificial seeds has become popular, where somatic embryos are encapsulated in a suitable matrix composed of sodium alginate, along with substances like mycorrhizae, herbicides, fungicides and insecticides. The technique involved in the production of artificial seeds is based on cellular totipotency and somatic embryogenesis.

An artificial seed is a bead of gel containing a somatic embryo (or shoot bud) and the nutrients, growth regulators, antibiotic, etc. needed for the development of a complete plantlet. Artificial seeds may be produced using one of the following two ways: desiccated systems and hydrated systems. In the desiccated systems the somatic embryos (SEs) are first hardened to withstand desiccation and then are encapsulated.

In the hydrated systems, the beads become hardened as calcium alginate is formed, after about 20-30 minutes the artificial seeds are removed, washed with water and used for planting. Hydrated artificial seeds become dry rapidly in the open air. Therefore, hydrated artificial seeds have to be planted soon after they are produced.

In India, this technique of synthetic seeds is being done for sandalwood and mulberry at BARC (Bhaba Atomic Research Centre), Mumbai.

Advantages:

1. They can be directly sown in the soil like natural seeds,

2. They can be stored upto a year without loss of viability,

3. They are easy to handle, and useful as units of delivery.

The only disadvantage of artificial seeds is the high cost of their production.

Practical Applications of Plant Tissue Culture

The use of plant cells to generate useful products and/or services constitutes plant biotechnology. In plant biotechnology, the useful product is a plantlet. The plantlets are used for the following purposes:s

1. Rapid Clonal Propagation:

   A clone is a group of individuals or cells derived from a single parent individual or cell through asexual reproduction. All the cells in callus or suspension culture are derived from a single explant by mitotic division. Therefore, all plantlets regenerated from a callus/suspension culture generally have the same genotype and constitute a clone. These plantlets are used for rapid clonal propagation. This is done in oil palm.

2. Somaclonal Variation:

   Genetic variation present among plant cells of a culture is called somaclonal variation. The term somaclonal variation is also used for the genetic variation present in plants regenerated from a single culture. This variation has been used to develop several useful varieties.

3. Transgenic plants:

   A gene that is transferred into an organism by genetic engineering is known as transgene. An organism that contains and expresses a transgene is called transgenic organism. The transgenes can be introduced into individual plant cells. The plantlets can be regenerated from these cells. These plantlets give rise to the highly valuable transgenic plants.

4. Induction and selection of mutations:

   Mutagens are added to single cell liquid cultures for induction of mutations. The cells are washed and transferred to solid culture for raising mutant plants. Useful mutants are selected for further breeding. Tolerance to stress like pollutants, toxins, salts, drought, flooding, etc. can also be obtained by providing them in culture medium in increasing dosage. The surviving healthy cells are taken to solid medium for raising resistant plants.

5. Resistance to weedicides:

   It is similar to induction of mutations. Weedicides are added to culture initially in very small concentrations. Dosage is increased in subsequent cultures till the desired level of resistance is obtained. The resistant cells are then regenerated to form plantlets and plants.

Explant Culture

Hairy roots production is carried out through the plant tissue culture technique in study plant metabolic processes or to manufacture precious secondary metabolites or recombinant proteins, often with plant genetic engineering. Hairy root culture is also called as transformed root culture from naturally occurring Gram negative soil bacterium Agrobacterium rhizogenes that contains root inducing plasmids (Ri plasmids). It infect roots of dicot and some monocot plants cause them to produce the opines which is a type of unusual amino acids (octopine, agropine, nopaline, mannopine, and cucumopine). Such opines are used by the bacterium as a carbon, nitrogen and energy source and to grow abnormally. Transformed roots are morphologically different from normal roots in that they Hairy root culture.

Hairy Root Culture was developed as the innovative path for bulky production of secondary metabolite, phytochemicals production. Thus this technique is of massive significance to develop large amount of roots and secondary metabolites in short time to continuous supply of improved value products.

Hairy Roots

Hairy roots production is carried out through the plant tissue culture technique in study plant metabolic processes or to manufacture precious secondary metabolites or recombinant proteins, often with plant genetic engineering. Hairy root culture is also called as transformed root culture from naturally occurring Gram negative soil bacterium Agrobacterium rhizogenes that contains root inducing plasmids (Ri plasmids). It infect roots of dicot and some monocot plants cause them to produce the opines which is a type of unusual amino acids (octopine, agropine, nopaline, mannopine, and cucumopine). Such opines are used by the bacterium as a carbon, nitrogen and energy source and to grow abnormally. Transformed roots are morphologically different from normal roots in that they much more branched and have much lateral meristematic growth, which will lead to higher biomass.

The abnormal roots are easy to culture in artificial media without hormone, and they are neoplastic in nature, with hazy growth. The hairy roots produced by Agrobacterium rhizogenes infection have a high growth rate as well as genetic and biochemical makeup. Hairy root culture is a kind of plant tissue culture that is use to study metabolic processes of plants, secondary metabolites production, production of recombinant proteins, plant genetic engineering, phytoremediation, artificial seed production, biofortification, biopharmaceuticals.

Perspectives of Hairy Root Culture

More Genotype and Phenotype Stability

Hairy root exhibits a high degree of chromosomal stability over prolonged culture period. The stability of hairy roots is an important advantage for both research and large scale industrial application. Stability demonstrated in terms of growth characteristics, DNA analysis, gene expression and secondary metabolites production.

High levels of secondary metabolite production

The common role of secondary metabolites in plants is defense mechanisms against their predators. Secondary metabolites are used as pharmaceuticals, agrochemicals, flavor, faineances, pesticides etc. The extraction of secondary metabolite from plants can not be economically synthesized due to its complex structure. Hence the use of hairy roots for the synthesis of secondary metabolites in larger quantity. Hairy roots shows high genetic stability as well fast growth rate on hormone free medium. Hairy root culture is unique tool for synthesis of high value secondary metabolite production and also valuable in studying secondary metabolite pathways such as follows:

- L-DOPA: A precursor of catecholamines, an important neurotransmitter used in the treatment of Parkinson's disease

- Shikonin: Use as an anti-ulcer agent and anti-bacterial

- Anthraquinone: used for dyes and medicinal purpose

- Opiate alkaloids: mostly codeine and morphine alkaloid use in medical purposes

- Berberine: an alkaloid with medicinal uses for cholera and bacterial dysentery

- Valepotriates: used as a sedative

- Ginsenosides: for medicinal purposes

- Rosmarinic acid: Use for medicinal purposes, antiviral and other suppression of endotoxin shock

- Quinine: for malaria

- Cardioactive or Cardenolides glycosides: To cure of heart disease.

Metabolic Studies using Phytoremediation

Environmental pollution is a global problem that is aggressive to all life forms including humans, animals and environment. The cost of cleaning up contaminated sites is very high. Genetically transformed hairy roots offer many practical advantages in experimental studies, such as ease of initiation, culture, and maintenance, indefinite propagation of material derived from the same parent plant, and genotypic and phenotypic stability. In phytoremediation use of plants for removal of environmental pollutants due to its low cost and safety of implementation. Transgenic plant roots make direct contact with pollutants in contaminated water or soil, for remediation of toxic substances and phytomining research .The hairy roots technology is an excellent platform for studying numerous aspects encompassing phytoremediation, xenobiotic biotransformation and degradation in plants, and for determining the responses of plant tissues to toxic heavy metals because hairy roots can be grown in large mass in culture media in a controlled environment and can therefore be subjected to various physiological assays. Also, these transformed roots are amenable to genetic operation and may make easy the categorization of genes that influence the phytoremediation capacity of plants. Thus, hairy roots offer a good prospect for the primary evaluation of transgenic efficacy in phytoremediation.

Production of Artificial Seed

Production of synthetic seeds is helpful in vitro plant propagation technology, because it having many useful advantages over a commercial scale for the propagation of a variety of crop plants. These tools provide important technique for production of synthetic seeds for conversion of plantlets under in vitro and in vivo circumstances. This technology is valuable for multiplication and conservation of best agricultural and endangered medicinal plant species, which are difficult to regenerate from natural seeds conventional methods. This technology developed in diverse cost-effectively significant plant species such as forage legumes, vegetable crops, industrially important crops, cereals, spices, fruit crops, plantation crops, medicinal plants, ornamental plants, orchids and wood yielding forest trees etc.

In artificial seed production experiments, the effect of concentrations of both gel matrix and time of exposure to calcium chloride on encapsulation of in vitro regenerated hairy roots with sodium alginate must be determined. Hairy roots in uniform size were taken and blot dried using filter paper and then mixed properly with the sodium alginate which prepared in distilled water. To absorb phenols and other compounds activated charcoal (0.2%) was added to the matrix to encapsulated hairy roots. The percentage of survival and conversion to plantlets were recorded after 60 days of storing.

Biofortification

Biotransformation is a process of breeding crops with higher levels of proteins, minerals, vitamins, and fat content micro-nutrients in crops. In such a way rising the dietary substance of the edible portion of plant foods are to levels that consistently exceed the average content. This can be done either through conventional selective breeding or through genetic engineering. For example Wheat variety Atlas 66, this has high protein content. Other examples of plants are rice, carrot, spinach etc.

Deficiencies of iron, zinc, and vitamin A influence in excess of one-half of the world's population. Progress has been made to control micronutrient deficiencies throughout food fortification and supplementation, but new approaches are needed, especially to reach the rural poor. Scientific evidence shows this is technically feasible without compromising agronomic productivity. Predictive cost-benefit analyses too maintain biofortification as individual important in the armamentarium for controlling micronutrient deficiencies.

Green Factories for Biopharmaceuticals

Biopharmaceuticals are high value therapeutic proteins that propose immense consequence in the treatment of various diseases like heart attack, cancer, diabetes, strokes, hemophilia and anemia. Commercial production of therapeutic proteins was done by using bacteria and mammalian cell cultures. There are some disadvantages related to the cost, scalability, safety and authenticity of protein related to the maintaining aseptic conditions. Plants are one of the most powerful alternatives for the bioproduction platforms because of their economic and safety advantages over the traditional method or commercial methods.

Bioaccumulation of Heavy Metals

Hairy roots are a convenient experimental tool for investigating the interactions between plant cells and metal ions. Hairy roots are proficient of collect heavy metals and we can investigate heavy metal tolerance in plants. Hairy roots also have potential for biological synthesis of quantum dot nanocrystals .Metal accumulation play important role in to understand plant biology, physiology and metabolism. Accumulation of heavy metals also shows some detrimental effect on cellular functions of plants.

Metals such as zinc, cadmium, nickel, manganese, cobalt, copper, are released into the environment from a range of sources of industries and agricultural field such as industrial effluent, sewage sludge, mining, military operations, fertilizers, and fossil fuel combustion. Soils and water metal contamination represents a risk to humans health, animals, ecosystems and also reduce soil fertility and crop yield in agricultural. Low-cost strategies are required for removing or. Dropping the bioavailability of metals from soil, environment, rivers and lakes, groundwater and sediments is very costly. Hence eco-friendly and low cost strategies are employed by production of hairy roots. The ability of plants to uptake and to detoxify metals can be supplementary exploited for the production of nanocrystals in living plant tissues.The nanoparticles of Au and Ag with dimensions 1 to 200 nm have been produced using a wide range of plant species ; the isolated nanoparticles play important role in medicine, chemical analyses, catalysis, biosensors, DNA detection, electrodes and electrical coatings.

Hyperhydricity

In general terms, swelling/thickening of the tissue, just like callus, is called vitrification or hyperhydricity. In many species, vitrification may be represented by symptoms not visible to the naked eye. It is a physiological malformation that results in excessive hydration, low lignification, poorly developed vascular bundles, impaired stomatal function, and reduced mechanical strength of tissue culture-generated plants. In vitrification, tissue becomes water soaked and translucent, which is mainly caused by excessive water uptake. It is usually controlled by changing agar concentration or source. The consequence is poor regeneration of such plants without intensive greenhouse acclimation for outdoor growth. Additionally, it may also lead to leaf-tip and bud necrosis in some cases, which often leads to loss of apical dominance in the shoots. In general, the main symptoms of hyperhydricity are translucent characteristics signified by a shortage of chlorophyll and high water content. Specifically, the presence of a thin or lack of a cuticular layer, reduced number of palisade cells, irregular stomata, less developed cell wall, and large intracellular spaces in the mesophyll cell layer have been described as some of the anatomic changes associated with hyperhydricity.

Main Causes of Hyperhydricity in Plant Tissue Culture

- Oxidative stresses as a result of high salt concentration.

- The type of explants utilized.

- The concentrations of microelement and hormonal imbalances.

- Low light intensity.

- High relative humidity.

- Gas accumulation in the atmosphere of the jar.

- Length of time intervals between subcultures.

- Number of subcultures.

- Concentration and type of gelling agent.

- High ammonium concentration.

- Evident in liquid culture-grown plants or when there is a low concentration of gelling agent.

Control of Hyperhydricity

- Monitoring the modified atmosphere of the culture vessels.

- Adjusting the relative humidity in the vessel.

- Use of gas-permeable membranes to increase exchange of water vapor and other gases such as ethylene with the surrounding environment.

- Use of higher concentration of a gelling agent to reduce the risk of hyperhydricity.

- Addition of agar hydrolysate.

- Use of growth retardants and osmotic agents.

- Use of bottom cooling, which allows water to condense on the medium.

- Use of cytokinin-meta-topolin (6-(3-hydroxybenzylamino)purine).

- Combination of lower cytokinin content and ammonium nitrate in the media.

- Use of nitrate or glutamine as the sole nitrogen source.

- Decreasing the ratio of ammonium: nitrate in the medium.

References

- The-difference-between-cultured-plant-cells-and-plant-stem-cells: annmariegianni.com, Retrieved 14 July 2018

- Plant-tissue-1941: maximumyield.com, Retrieved 12 March 2018

- Plant-tissue-culture-environmental-condition-methods-types-dey-1: linkedin.com, Retrieved 14 April 2018

- Plant-tissues-and-organs: lumenlearning.com, Retrieved 20 May 2018

- Hairy-root-culture-a-promising-approach-in-biotransformation: imedpub.com, Retrieved 18 June 2018

- Hyperhydricity, agricultural-and-biological-sciences: sciencedirect.com, Retrieved 27 March 2018

Chapter 3

Micropropagation

Micropropagation is a vegetative propagation technique for the production of a number of pathogen-free and genetically-superior transplants in a limited space in a limited time. A detailed study of the fundamental aspects of micropropagation has been provided in this chapter, particularly meristem, callus, embryo, suspension and protoplast culture.

Micropropagation is a technique that manipulates small quantities of axenic plant material, ranging from single cells to stem segments, under conditions favourable to the formation of new plants. It has proven to be the most efficient and cost-effective method of propagating large numbers of clonal offspring for many agronomic crops, including both herbaceous and woody perennial species. Older and simpler techniques of cloning plants (cuttings, grafting, and division of parent stock material) are limited by seasonal constraints and the natural formation of new plant structures. Micro propagation, on the other hand, allows the year-round production of new plants at rates significantly higher than that achievable by all other methods. The plants produced are genetically uniform, vigorous, and free from associations with other organisms, an attribute particularly useful for the culture of underwater grasses where contaminating organisms can dominate other types of production systems.

Methods and Approaches

Explants and their Surface Disinfection

Small pieces of plants (explants) are used as source material to establish cells and tissues in vitro. All operations involving the handling of explants and their culture are carried out in an axenic (aseptic; sterile) environment under defined conditions, including a basal culture medium of known composition with specific types and concentrations of plant growth regulators, controlled light, temperature and relative humidity, in culture room(s) or growth cabinet(s). The disinfection of explants before culture is essential to remove surface contaminants such as bacteria and fungal spores. Surface disinfection must be efficient to remove contaminants, with minimal damage to plant cells.

Culture Media and their Preparation

Culture media contain microelements, microelements, vitamins, other organic com-ponents (e.g. amino acids), plant growth regulators, gelling agents (if semisolid)and sucrose. Gelling agents are omitted for liquid media. The composition of the culture medium depends upon the plant species, the explants, and the aim of the experiments. In general, certain standard media are used for most plants, but some modifications may be required to achieve genotype-specific and stage-depend entoptimizations, by manipulating the concentrations of growth regulators, or by the addition of specific components to the culture medium. Commercially available ready-made powdered medium or stock solutions can be used for the preparation of culture media. A range of culture media of different formulations, and plant growth regulators are supplied by companies such as Duchefa and Sigma-Aldrich.Murashige and Skoog medium (MS) is used most extensively.

Stages of Micropropagation

The following distinct stages are recognized for the micro propagation of most plants:

- Stage I: Establishment of axenic cultures – introduction of the surface disinfected explants into culture, followed by initiation of shoot growth. The objective of this stage is to place selected explants into culture, avoiding contamination and providing an environment that promotes shoot production. Depending on the type of explant, shoot formation may be initiated from apical and axillary buds (pre-existing meristems), from adventitious meristems that originate on excised shoots, leaves, bulb scales, flower stems or cotyledons (direct organogenesis), or from callus that develops at the cut surfaces of explants (indirect organogenesis). Usually 4–6 weeks are required to complete this stage and to generate explants that are ready to be moved to Stage II. Some woody plants may take up to 12 months to complete Stage I, termed 'stabilization'. A culture is stabilized when explants produce a constant number of normal shoots after subculture.

- Stage II: Multiplication – shoot proliferation and multiple shoot production. At this stage, each explant has expanded into a cluster of small shoots. Multiple shoots are separated and transplanted to new culture medium. Shoots are subculture every 2–8 weeks. Material may be sub cultured several times to new medium to maximise the quantity of shoots produced.

- Stage III: Root formation – shoot elongation and rooting. The rooting stage pre-pares the regenerated plants for transplanting from in vitro to ex vitro conditions in controlled environment rooms, in the glasshouse and, later, to their ultimate loca-tion. This stage may involve not only rooting of shoots, but also conditioning of the plants to increase their potential for acclimatization and survival during trans-planting. The induction of adventitious roots may be achieved either in vitro or exvitro in the presence of auxins. The main advantage of ex vitro compared to in vitro rooting is that root damage during transfer to soil is less likely to occur.The rates of root production are often greater and root quality is optimized when rooting occurs ex vitro.

- Stage IV: Acclimatization – transfer of regenerated plants to soil under natural environmental condition. Transplantation of in vitro-derived plants to soil isoften characterized by lower survival rates. Before transfer of soil-rooted plant sto their final environment, they must be acclimatized in a controlled environment room or in the glasshouse. Plants transferred from in vitro to exvitro conditions, undergo gradual modification of leaf anatomy and morphology, and their stomata begin to function (the stomata are usually open when the plants are in culture). Plants also form a protective epicuticular wax layer over the surface of their leaves. Regenerated plants gradually become adapted to survival in their new environment.

Techniques of Micropropagation

Cultures of Apical and Axillary Buds

Currently, the most frequently used micro propagation method for commercial mass production of plants utilizes axillary shoot proliferation from isolated apical or axillary buds under the influence of a relatively high concentration of cytokinin.

In this procedure, the shoot apical or axillary buds contain several developing leaf primordia. Typically, the explants are 3–4 mm in diameter and 2 cm in length. Development in vitro is regulated to support the growth of shoots, without adven-titious regeneration.

Meristem and Single- or Multiple-node Cultures (Shoot Cultures)

Meristems are groups of undifferentiated cells that are established during plantembryogenesis. Meristems continuously produce new cells which undergo differentiation into tissues and the initiation of new organs, providing the basic structure of the plant body. Shoot meristem culture is a technique in which adome-shaped portion of the meristematic region of the stem tip is dissected from as elected donor plant and incubated on culture medium. Each dissected meristem comprises the apical dome with a limited number of the youngest leaf primordia,and excludes any differentiated provascular or vascular tissues. A major advan-tage of working with meristems is the high probability of excluding pathogenicorganisms, present in the donor plant, from cultures. The culture conditions are controlled to allow only organized outgrowth of the apex directly into a shoot,without the formation of any adventitious organs, ensuring the genetic stability of the regenerated plants.

The single-or multiple-node technique involves production of shoots from cul-tured stem segments, bearing one or more lateral buds, positioned horizontally or vertically on the culture medium c. Axillary shoot proliferation from the buds in the leaf axils is initiated by a relatively high cytokinin concentrationd. Meristem and node cultures are the most reliable for micro propagation to produce true-to-type plants.

Adventitious Shoot Formation

Adventitious shoot formation is one of the plant regeneration pathways in vitro,and is employed extensively in plant biotechnology for micro propagation and genetic transformation, as well as for studying plant development. Adventitious meris-tems develop de novo and in vitro they may arise directly on stems, roots or leaf explants, often after wounding or under the influence of ex-ogenous growth regu-lators (direct organogenesis). Cytokinins are often applied to stem, shoot or leaf cuttings to promote adventitious bud and shoot formation. Adventitious buds and shoots usually develop near existing vascular tissues enabling the connection with vascular tissue to be observed. Adventitious organs sometimes also originate in callus that forms at the cut surface of explants (indirect organogenesis). Somaclonal variation, which may be useful or detrimental, may occur during adventitious shoot regeneration.

Somatic Embryogenesis

Somatic embryogenesis was defined by Emons as the development fromsomatic cells of structures that follow a histodifferentiation pattern which leads to a body pattern resembling that of zygotic embryos. This process occurs naturally in some plant species and can be also induced in vitro in others species. There is considerable information available on in vitro plant regeneration from somatic cells by somatic embryogenesis. Somatic embryogenesis may occur directly from cellsor organized tissues in explants or indirectly through an intermediate callus stage.

It has been confirmed in many species that the auxins 2,4-D and NAA, in the correct concentra-tions, play a key role in the induction of somatic embryogenesis.Application of the cytokinins,

BAP or kinetin, may enhance plant regeneration from somatic embryos after the callus or somatic embryos have been induced by auxin treatment. However, in some species (such as Abies alba) cytokinins on their own induce somatic embryogenesis.

Advantages of Micropropagation

1. Requires relatively small growing space.

2. The technique of micro propagation is applied with the objective of enhancing the rate of multiplication. Through tissue culture over a million plants can be grown from a small, even microscopic, piece of plant tissue within 12 months.

3. Shoot multiplication usually has a short cycle (2-6 weeks) and each cycle results in logarithmic increase in number of shoots.

4. Tissue culture gives propagules such as minitubers or microcorms for plant multiplication throughout the irrespective of the season.

5. The small size of propagules and their ability to proliferate in a soil free environment facilitate their storage on a large scale ability to proliferate in a soil free environment facilitate their storage on a large scale and also allows their large scale dissemination by suitable means of transport across international boundries.

6. Stocks of germplasm can be main for many years.

7. Pathogen free plants can be raised and maintained economically.

8. Clonal propagation in dioecious is extremely important since seed progently yields 50% females and 50% male plants of one sex are desired commercially. For example, male plants of asparagus officinalis are more valuable then female plants. Propagation by stem cutting in male asparagus is not successful but can be achieved by tissue culture.

Limitations of Micropropagation

1. Sophisticated facilities are required.

2. Demands greater skill in handling and maintenance than conventional techniques.

3. Shoot-tip derived plants may show genetic instability, E.g. 6-35% of banana clones developed through shoot tip culture show morphological variation.

Shoot-tip and Meristem Culture

Shoot tip culture may be described as the culture of terminal (0.1-1.0 mm) portion of a shoot comprising the meristem (0.05-0.1 mm) together with primordial and developing leaves and adjacent stem tissue. On the other hand, meristem culture is the in vitro culture of a generally shiny special dome-like structure measuring less than 0.1 mm in length and only one or two pairs of the youngest leaf primordia, most often excised from the shoot apex.

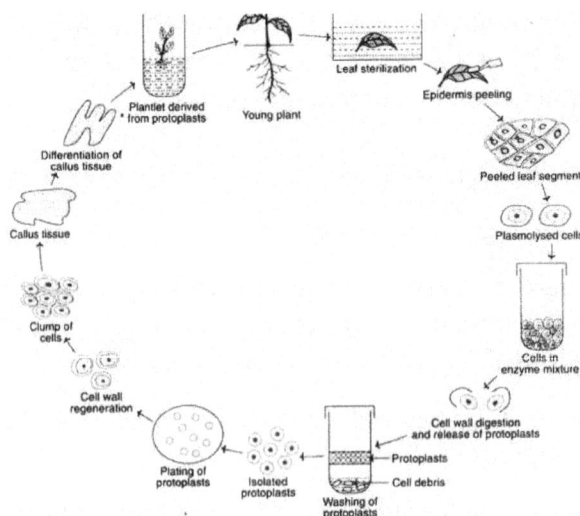

Principle

The excised shoot tips and meristem can be cultured aseptically on agar solidified simple nutrient medium or on paper bridges dipping into liquid medium and under the appropriate condition will grow out directly into a small leafy shoot or multiple shoots. Alternatively the meristem may form a small callus at its cut case on which a large number of shoot primordia will develop.

These shoot primordia grow out into multiple shoots. Once the shoots have been grown directly from the excised shoot tip or meristem, they can be propagated further by nodal cuttings. This process involves separating the shoot into small segments each containing one node. The axillary bud on each segment will grow out in culture to form yet another shoot.

The excised stem tips of orchids in culture proliferate to form callus from which some organised juvenile structures known as proto-corm develop. When the proto-corms are separated and cultured to fresh medium, they develop into normal plants. The stem tips of Cascuta reflexa in culture can be induced to flower when they are maintained in the dark.

Exogenously supplied cytokinins in the nutrient medium plays a major role for the development of a leafy shoot or multiple shoots from meristem or shoot tip. Generally, high cytokinin and low auxin are used in combination for the culture of shoot tip or meristem.

Addition of adenine sulfate in the nutrient medium also induces the shoot tip multiplication in some cases. BAP is the most effective cytokinin commonly used in shoot tip or meristem culture. Similarly, NAA is the most effective auxin used in shoot tip culture. Coconut milk and gibberellic acid are also equally effective for the growth of shoot apices in some cases.

Protocol

- Remove the young twigs from a healthy plant. Cut the tip (1 cm) portion of the twig.

- Surface sterilize the shoot apices by incubation in a sodium hypochlorite solution (1% available chlorine) for 10 minutes. The ex- plants are thoroughly rinsed 4 times in sterile distilled water.

- Transfer each explants to a sterilized petri dish.

- Remove the outer leaves from each shoot apices with a pair of jeweler's forceps. This lessens the possibility of cutting into the softer underlying tissues.

- After the removal of all outer leaves, the apex is exposed. Cut off the ultimate apex with the help of scalpel and transfer only those less than 1 mm in length to the surface of the agar medium or to the surface of filter-paper Bridge. Flame the neck of the culture tube before and after the transfer of the excised tips. Binocular dissecting microscope can be used for cutting the true meristem or shoot tip perfectly.

- Incubate the culture under 16hrs light at 25°C.

- As soon as the growing single leafy shoot or multiple shoots obtained from single shoot tip or meristem, develop root, transfer them to hormone free medium.

- The plantlets formed by this way are later transferred to pots containing compost and kept under greenhouse conditions.

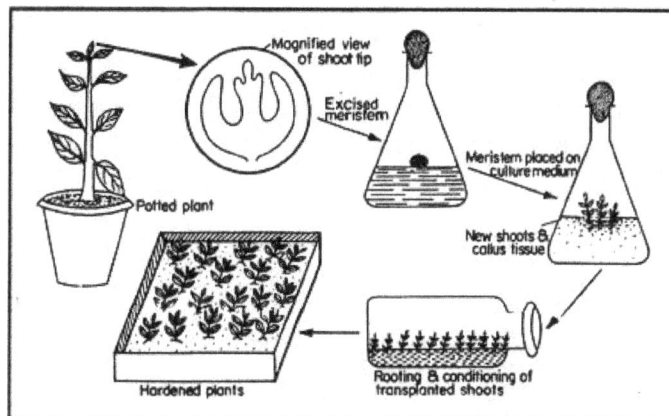

Importance of Shoot Tip/Meristem Culture

The uses of shoot tips and meristem in tissue culture are very varied and include mainly:

1. Virus eradication,

2. Micro-propagation and

3. Storage of genetic resources.

Virus Eradication

Many important plants contain systemic viruses which substantially reduce their potential yield and quality. It is, therefore, important to produce virus free stocks which can be multiplied. Generally, highly meristematic tissue of a virus infected plant remains free from virus due to their fast mitotic activity. Therefore, shoot tips and meristems of a virus infected plant are the ideal explants to produce a virus free stock.

This technique is also valuable for the maintenance of carefully defined stocks of specific varieties and cultivars in disease Free State. The size of the meristem explant is critical for virus eradication. Often so called meristem tip cultures have failed to eliminate virus infection because the explant contains shoot apices with vascular tissue instead of true meristem.

This technique, combined with heat treatment (thermotherapy) or chemical treatment (chemotherapy) has proved to be very effective in virus eradication. Heat treatment is done by placing an actively growing plant in a thermotherapy chamber. Over a period of two weeks the temperature is increased to 38°C inside the chamber and the plants are maintained at this temperature for two months.

After that period, the apical meristem is excised, surface sterilized and transferred aseptically to agar medium. Using this technique 85% to 90% virus free plants have been obtained. Without heat treatment, shoot tips or meristem can be grown on chemotherapeutants added medium for virus eradication. Commonly used chemotherapeutants are 2, 4-D, mela-chife green, thiouracil etc.

Shoot tip or meristem culture has enormous horticultural value e.g. in the production of plants for the cut flower industry when stock plants of registered lines must be maintained in as near-perfect condition as possible. Any infection by virus that affect the growth or physical characteristics of size and shape of flowers, is obviously very serious problem from commercial point of view.

Meristem culture technique to clean up the stocks could, therefore, avert a commercial disaster. Similarly, in the agricultural world, the production or yield of a crop can fall dramatically as a result of a viral infection and render that particular variety no longer salable or commercial value. Meristem culture could be of value in restoring the original properties of the variety by removing the infection.

Micro-propagation

A sexual or vegetative propagation of whole plants using tissue culture techniques is referred to as micro-propagation. Shoot tip or meristem culture of many plant species can successfully be used for micro-propagation.

- Storage of Genetic Resources

  Many plants produce seeds that are highly heterozygous in nature or that is recalcitrant. Such seeds are not accepted for storing genetic resources. So, the meristem from such plants can be stored in vitro. Besides the above-mentioned uses of shoot tip or meristem culture, it can also be utilized in various important fields of plant science such as:

- Shoot Tip or Meristem Culture and Plant Breeding:

In many plant breeding experiments, the hybrid plants produce abortive seeds or nonviable seeds. As a result, it makes a barrier to crossibility in plants where non-viable seeds are unable to develop into mature hybrid plants. Shoot tip or meristem from such hybrid plant can be cultured to speed up breeding programme.

- Propagation of Haploid Plants:

Haploid plants derived from anther or pollen culture always remain sterile unless and until they are made homozygous diploid. Meristem or shoot tip culture of haploid plants can be used for their propagation from which detailed genetic analysis can be done on the basis of morphological characters and biochemical assay.

- Meristem or Shoot Tip Culture and Quarantine:

There are some strict regulations concerning the international movement of vegetative plant material. After thoroughly checking, the plant materials may be rejected by quarantine authority if the plant material carries some diseases.

Plantlets derived from shoot tip or meristem culture are easily accepted by the quarantine authority for international exchange without any checking. Therefore, using this technique, crop plants can be easily exchanged in crop improvement programmes that are based on materials from different parts of the world.

Callus Culture

In callus culture, cell division in explant forms a callus. Callus is irregular unorganised and undifferentiated mass of actively dividing cells. Darkness and solid medium gelled by agar stimulates callus formation. The medium ordinarily contains the auxin, 2,4-D, (2, 4- dichlorophenoxy acetic acid) and often a cytokinin like BAP (Benzyl aminopurine). Both are growth regulators. This stimulates cell divison in explant. Callus is obtained within 2-3 weeks.

Higher plant body is multicellular and is made up of highly organised and differentiated structures like stem, leaf, root, etc. different tissue system present in different organs function in a highly coordinated manner. Now, if the organised tissue are diverted into an unorganised proliferation mass of cell by any means, they will form the callus tissue.

In nature, sometimes callus or callus –like tissue is found to form to form in various part of intact plant either due to deep wound or due to some disease. Deep large wound in branches and trunks of intact plants results in the formation of soft mass of unorganised parenchymatous tissues which are rapidly formed on or below the injured surface of the organ concerned. Such callus tissue is known as wound callus. Wound callus is formed by the division of cambium tissues. They may also be formed by the same process from the parenchymatous cells of cortex, phloem and xylem rays. Callus like growth is also stimulated due to some disease caused by Agrobacterium tumefaciens, synchytrium endobioticum and virus, insects etc.

Such callus –like outgrowth is known as gall or tumour. But the callus in tissue culture is produce experimentally from the small excise portion called the explant of any living healthy plant. In culture, the excised plant tissue losses its structural integrity and changes completely to a rapidly proliferative unorganised mass of cells which is called the callus tissue.

Nutrient Medium of Callus Culture

Some standard media, such as, Murashige and Skoog's medium can be successfully used for callus culture. For initiation and maintaining callus kinetin is widely used in the medium.

For callus initiation usually an exogenous supply of hormone is required. But explants having cambial cells do not require a supply of hormone. According to hormone requirements callus culture may be of five types.

These are:

- Auxin requiring cultures,

- Cytokinin requiring cultures,

- Cultures requiring both auxin and cytokinin,

- Gibberellin requiring cultures. In some plants, such as tobacco, presence of gibberellin and N6 2 isopentenyl adenine in the medium favours callus growth. But gibberellin inhibits growth of callus tissue in monocots,

- cultures requiring other natural extracts, such as, yeast extract, coconut milk, casein hydrolysate or tomato juice etc.

Methods of Callus Culture

Usually explants from suitable materials (such as, carrot root, potato or sweet potato tuber, stem of tobacco, hypocotyl and cotyledon of soya bean etc..) are taken. The explants is first surface sterilised with 1.6% sodium hypochlorite solution or 0.1% mercuric chloride solution or 1% aqueous solution of bromine.

Then the inner uncontaminated tissue is excised. If the excised tissue (such as, root, hypocotyl, cotyledon etc.) is taken from a seedling then the seed before germination is surface sterilised and allowed to germinate under aseptic conditions.

Development of a Callus Culture

Callus formation from an explants occurs in three stages:

- Induction stage:

  Metabolism is stimulated and the cells prepare to divide. Cell size remains unchanged.

- Cell division stage:

  Cells divide actively and the cell size decreases. Cell division is mainly periclinal and occurs towards the periphery giving rise to wound cambial cells.

- Differentiation:

  Cells differentiate by expansion and maturation. Rapidly growing calluses are more or less alike but as the growth rate decreases the calluses show their characteristic structures and forms. But all calluses have some similarities. They all contain nodular or sheet meristems in groups or scattered throughout the tissue.

Vascular nodules or meristemoids are formed in a callus from small groups of meristematic cells. The nodules may not differentiate or may form root or shoot primordia or embryos.

These meristemoids resemble vascular bundles and consist of xylem, phloem and cambium. Measurement of growth in a callus culture is based on fresh weight or dry weight or cell number counts. The callus may be weighed directly under aseptic conditions.

Suspension Culture

Suspension culture is a type of culture in which single cells or small aggregates of cells multiply while suspended in agitated liquid medium. It is also referred to as cell culture or cell suspension culture. Establishment of single cell cultures provides an excellent opportunity to investigate the properties and potentialities of plant cells. Such systems contribute to our understanding of the interrelationships and complementary influences of cells in multicellular organisms. Many plant biotechnologists recognized the merits of applying cell cultures over an intact organ or whole plant cultures to synthesize natural products.

Principles of Cell Suspension Culture

1. The basic principle of single cell culture is the isolation of large number of intact living cells and cultures them on a suitable nutrient medium for their requisite growth and development.

2. Callus proliferates as an unorganized mass of cells. So it is very difficult to follow many cellular events during its growth and developmental phases. To overcome such limitations, the cultivation of free cells as well as small cell aggregates in a chemically defined liquid medium as a suspension was initiated.

3. In culture, the single cells divide re divide to form a callus tissue. Such callus tissue also retains the capacity to regenerate the plantlets through organogenesis and embryogenesis.

4. To achieve an ideal cell suspension, most commonly a friable callus is transferred to agitated liquid medium where it breaks up and readily disperses.

5. After eliminating the large callus pieces, only single cells and small cell aggregates are again transferred to fresh medium and after two or three weeks a suspension of actively growing cells are produced.

6. This suspension can then be propagated by regular sub-culture of an aliquot to fresh medium.

7. Ideally suspension culture should consist of only single cells which are physiologically and biochemically uniform.

8. Movement of cells in relation to nutrient medium facilitates gaseous exchange, removes any polarity of the cells due to gravity and eliminates the nutrient gradients within the medium and at the surface of the cells.

Methodology

1. To achieve an ideal cell suspension most commonly a friable callus is transferred to agitated liquid medium where it breaks up and readily disperses.

2. After eliminating the large cellular pieces, only single cells and small cell aggregates are again transferred to fresh medium and after 2 or 3 weeks a suspension of actively growing cells are produced.

Isolation of Single Cells

• From Plant Organs

The most suitable material for the isolation of single cells is the leaf tissue, since a more or less homogenous population of cells in the leaves offer good material for raising defined and controlled large scale cell cultures.

Two important methods to isolate single cells from leaf are:

• Mechanical Method

• Enzymatic Method

From Cultured Tissues

The most widely applied approach is to obtain a single cell system from cultured tissues.

From Plant Organs

Mechanical Method

Gnanam and Kulandaivelu (1969) developed a procedure which has since been successfully used to isolate mesophyll cells active in photosynthesis and respiration, from mature leaves of several species of dicots and monocots including the grasses.

The procedure involves:

1. Mild maceration of 10g leaves in 40ml of the grinding medium (20µ mol. Sucrose, 10µ mol MgCl2, 20µ mol tris HCl buffer, pH 7.8) with a mortor and pestle.

2. The homogenate obtained is passed through two layers of muslin cloth and the cells thus released are washed by centrifugation at low speed using the medium.

The mechanical isolation of free parenchymatous cells can also be achieved on a large scale.

Enzymatic Method

In 1968 Takabe et al treated tobacco leaf tissue with the enzyme pectinase and obtained a large number of metabolically active cells. A point to note is that potassium dextran sulphate in the enzyme mixture improved the yield of free cells.

Isolation of single cells by the enzymatic method has been found convenient as it is possible to obtain high yields from preparations of spongy parenchyma with minimum damage or injury to the cells. This can be accomplished by providing osmotic protection to the cells while providing osmotic protection to the cells while the enzyme macerozyme degrades the middle lamella and cell wall of the parenchymatous tissue. Applying the enzymatic method to cereals has proven difficult since the mesophyll cells of these plants are apparently elongated with a number of interlocking constrictions, thereby preventing their isolation.

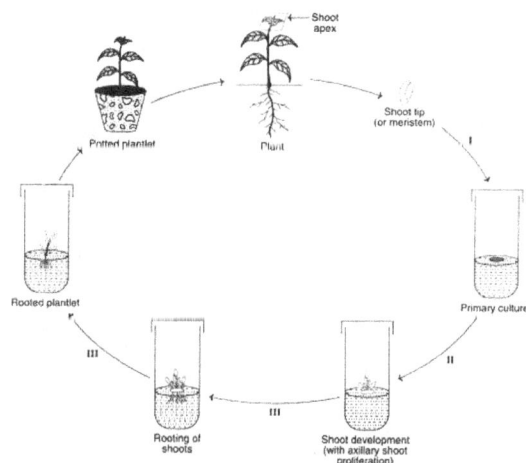

From Cultured Tissues

*   Raise sterile tissue culture plants and obtain callus from them.

- The callus is separated from an explant and transferred to a fresh medium of the same composition to enable it to build up a mass of tissue.

- Repeated subculture on an agar medium improves the friablity of a callus, a pre requisite for raising the fine cell suspension in a liquid medium.

- The pieces of undifferentiated and friable callus are transferred in a continuously agitated liquid medium dispersed in autoclaved flasks or other suitable vials.

- Agitation is done by placing the flasks on orbital platform shaker or suitable device.

- Movement of the culture medium exerts mild pressure on small pieces of tissues breaking them into free cells and small aggregates. Further it augments the gaseous exchange between the culture medium and the culture air and also ensures uniform distribution of cells in the medium.

- The period of incubation during which the suspension cultured is developed from callus tissue is usually called as the initiation passage.

- The concentration of auxins and cytokinins is often critical for the growth of cell suspension and the concentration of auxin and cytokinins used for callus culture is generally reduced for suspension culture.

- The cells in the cell suspension may vary in shapes and sizes. They may be oval, round, elongated or coiled, but thery are thin walled, even with the presence of other lignified, trachieds like elements.

Protocol

- Take 150/250 ml conical flask containing autoclaved 40/60 ml liquid medium.

- Transfer 3-4 pieces of pre-established callus tissue (appx. Wt. 1 gm each) from the culture tube using the spoon headed spatula to conical flasks.

- Flame the neck of conical flask, close the mouth of the flask with a piece of aluminium foil or a cotton plug. Cover the closure with a piece of brown paper.

- Place the flasks within the clamps of a rotary shaker moving at the 80 – 120 rpm.

- After 7 days, pour the contents of each flask through the sterilized sieve and collect the filtrate in a big sterilized container. The filtrate contains only free cells and cell aggregate.

- Allow the filtrate to settle for 10 – 15 min or centrifuge the filtrate at 500 – 1000 rpm and finally pour off the supernatant.

- Resuspend the residue cells in a requisite volume of fresh liquid medium and disperse the cell suspension equally in several sterilized flasks (150/250 ml). Place the flasks on shaker and allow the free cells and cell aggregates to grow.

- At the next subculture, repeat the previous steps but take only one-fifth of the residual cells as the inoculums and dispense equally in flasks and again place them on shaker.

- After 3- 4 subcultures, transfer 10ml of cell suspension from each flask into new flask containing 30ml fresh liquid medium.

- To prepare a growth curve of cells in suspension, transfer a definite number of cells measured accurately by a haemocytometer to a definite volume of liquid medium and incubate on shaker. Pipette out very little aliquot of cell suspension at short intervals of time (1 or 2 days interval) and count the cell number. Plot the cell count data of a passage on a graph and the curve will indicate the growth pattern of suspension culture.

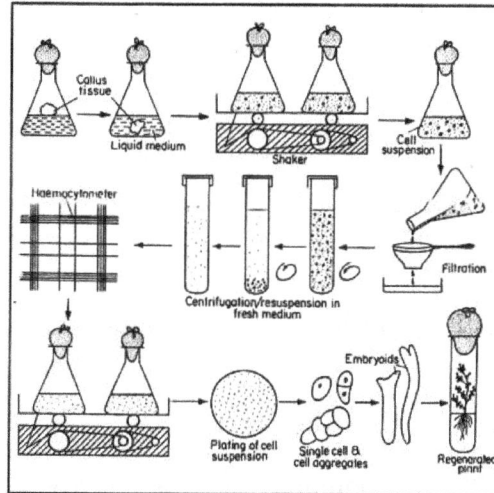

Different Categories of Cell Suspension Culture

Broadly speaking there are two types of suspension cultures:

1. Batch Culture

2. Continuous Culture

Batch Culture

Here the cell material grows in a finite volume of agitate liquid medium. For instance, cell material in 20 ml or 40 ml or 60 ml liquid medium in each passage constitute a batch culture.

Batch suspension cultures are most commonly maintained in conical flasks incubated on orbital platform shakers at the speed of 80 – 120 rpm.

Different types of batch culture are:

1. Slowly rotating cultures

2. Shake cultures

3. Spinning cultures

4. Stirred culture

Continuous Culture

The large culture vessel is kept dispersed continuously by bubbling sterile air through culture medium and the old liquid medium is continuously replaced by the fresh liquid medium (on depletion of some nutrients in the medium) to stabilize the physiological states of the growing cells.

Here nutrient depletion does not occur due to continuous flow of nutrient medium and the cells always remain in the steady state of active growth phase.

There are two types of continuous culture system:

1. Chemostates

2. Turbidostates

Importance of Cell Suspension Culture

- To obtain single cell clones.

- To study the morphological and biochemical changes during their growth and development phases.

- To understand the pathways of cellular metabolism.

- Single cell systems have a great potential for crop improvement.

- Free cells in cultures permit quick administration and withdrawal of diverse chemicals/ substances thereby making them easy targets for mutant selection.

- Cells which are in a population of cultured cells invariably show cytogenetical and metabolic variations depending on the stage of the growth cycle and culture (Lindsey and Yeoman (1985) Heterogeneity).

- To produce high yielding cultures as well as plants with superior agronomic traits.

- Single cells derived from medicinally important plants can be studied for the production of secondary metabolites like alkaloids, glycosides.

- For mutagenesis study. The mutagens can be added directly in the liquid medium. After the mutagen treatment, cells are plated on agar medium for the selection of mutant cell clones. The hope is that permanent changes in the DNA patterns of some of the cells would be achieved by such treatments.

Embryo Culture

In addition to root, shoot, and pollen culture, embryo culture has also been done for the production of haploid plants. Embryo culture is used for the recovery of plants from distinct crosses. Embryo culture is useful where embryo fails to develop due to degeneration of embryonic tissues.

It is being used extensively in the extraction of haploid barley (Hordeum vulgare) from the crosses H. vulgare x H. bulbosum. Embryo culture is also a routine technique employed in orchid propagation and in breeding of those species that show dormancy. Das and Barman (1992) developed the method of regeneration of tea shoots from embryo callus. The embryo callus produced somatic embryoids within 8 weeks of culture in the second medium which differentiated into buds after 2 weeks. Several shoots with 4-6 leaves developed after 16 weeks of culture.

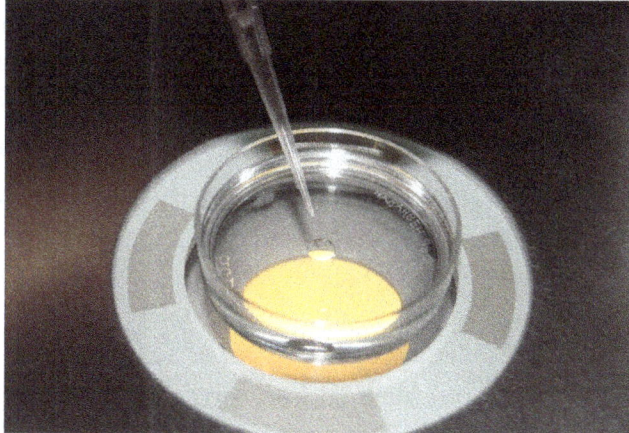

Technique of Embryo Culture

Surface Sterilization

Embryos of seed plants normally develop inside the ovule which in turn is covered by overies. Since they already exist in a sterile environment, disinfection of the embryo surface is unnecessary unless the seed coats are injured or systemic infection is present. Instead, mature seeds, entire ovule or fruits are surface sterilized. Surface sterilization is carried out by immersing the material in hypo chorine- containing commercial bleach (5-10% Clorox, 0.45% Sodium or Calcium hypochlorite) for 5- 10 min or ethanol (70-75%) for 5 min. A small amount (0.01-0.1%) of a surfactant may be added to disinfection solution. In case of infected seeds, the excised embryos may be immersed in 70% alcohol plus 5-10 min exposure to 2.6% sodium hypochlorite solution.

Excision of Embryo

Embryo excision operation is carried out aseptically in a laminar airflow hood. A steremicroscope equipped with cool-ray flurescent lamp is required for excision of small embryo. The commonly used dissecting tools are foreceps, dissecting needles, scalpels, razor blades and Pasteur pipettes. Mature embryo can be isolated with relative ease by splitting open the seeds. Soaking a hard coat seeds for few hours to a few days before sterilization makes its dissection easier. In case of embryos embedded in liquid endosperms, the incision is made at micropolar end of young ovule and pressure applied at opposite end to force the embryo out through the incision.

Embryo-endosperm Transplant

It is very difficult to culture embryo in vitro abort at very early stages of development because of lack of knowledge of nutritional requirements. The chances of development of immature or abortive embryos increases if they are surrounded by endosperm tissue excised from another seed of same species. Generally, endosperm older than the embryo by 5 days was more efficient as a nurse tissue than one of the same age as the embryo.

Nutritional Requirement

The nutritional requirements of an embryo during its development in vitro consisting two phase: a) Heterotropic phase- an early phase wherein the embryo is dependent and draws upon the endosperms and materal tissues and b) The autotrophic phase- a later phase in which the embryo is metabolically capable of synthesizing substances required for its growth, thus becoming fairly independent for nutrition.

The media constituents for in vitro growth of young or immature embryos also differ from those of mature embryos. This often necessitates the transfer of embryos from one medium to another for their orderly growth.

i) Mineral Salts:

Inorganic nutrients of MS, B5 and White's media with certain degree of modification are the most widely used basal media for embryo culture. Monnier (1978) modified the MS medium for immature embryo culture of Capsella which contains higher levels of potassium and calcium and reduced levels of ammonium (NH_4NO_3) and FeEDTA and double concentration of MS micronutrients.

ii) Carbohydrates:

Sucrose is the most commonly used source of energy for embryo culture. Addition of maltose, lactose, raffinose or mannitol may be required in embryo culture of some species. In some cases glucose is found to be better than sucrose. Mature embryos grow fairly well at low sucrose concentration but younger embryos require higher level of carbohydrates.

iii) Nitrogen and Vitamins:

Ammonium nitrate is better than KNO_3, $NaNO_3$ and $(NH_4)_2 HPO_4$. especially the presence of NH_4+ in the medium has been found essential for proper growth and differentiation of embryos. Various Amino acids and their amides like aspargine, glutamine, and casein hydrolysate have been widely used in embryo culture.

iv) Natural Plant Extract:

The coconut milk (CM) effectively stimulates the growth of excised young embryos of sugarcane, barley, tomato, carrot, Interspecific hybrids of Vigna and fern species. Van Overbeek et.al (1941) suggested that the coconut milk contains some ' Embryofactor' which presumably makes up for deficiencies of certain sugars, amino acids, growth hormones and other critical metabolites of the culture medium.

In addition to coconut milk, water extracts from dates, bananas, hydrolysate of wheat-gluten and tomato juice were also effective.

v) Growth Regulators:

Auxin and cytokinins are not generally used in embryo culture since they induce callus formation. At very low concentration GA promotes embryogenesis of young barley embryos without inducing precocious germination. ABA also has a similar effect on barley and Phaseolus embryos.

vi) PH of Medium:

Excised embryos grows well in a medium with a PH 5 to 7.5. Generally the medium PH is adjusted 0.5 units higher than the desired PH in order to compensate for uncontrollable change in its value during the autoclaving process.

vii) Incubation Conditions:

The embryo cultures are incubated at 25±2 °C. whereas in case of species to warm temperature requires 27-30 °C incubation temperature and species occurring in cold regions or seasons require incubation temperature of 17-22 °C.

Generally, an initial dark incubation (4 days) of embryo in culture is essential , following which they can grow to a mature stage even under continuous light regime.

Role of Suspensor in Embryo Culture

Suspensor is actively involved in embryo development. The suspensor is an ephemeral structures found at the radicular end of the proembryo and attains maximum development by the time embryo reaches globular stage. In cultures the presence of a suspensor is critical , particularly for the survival of young embryos. The requirement of the suspensor may be substituted by the addition of GA or ABA to the culture medium.

Protoplast Culture

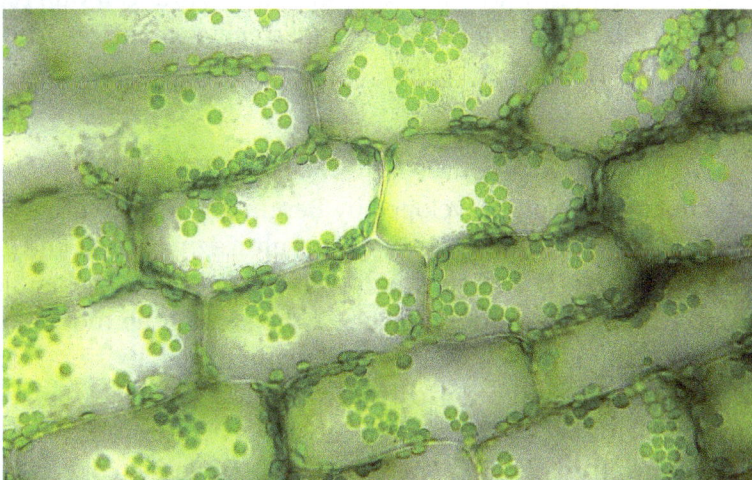

After viability test the protoplasts are cultured at a known density.
Different methods have been used for culturing the isolated protoplasts.

Suspension Culture

In this method protoplasts are suspended in a liquid medium with a suitable concentration of os moticum. Protoplasts at a density of 105/ml is generally plated on 25-50 ml of medium in an Erlenmeyer flask. The cultures are shaken slowly to provide sufficient aeration for growth. Sometimes shaking can cause bursting of protoplasts, so rpm of shaking required for a given species should be standardized. 2 ml suspension of protoplasts could be cultured in 25 ml flasks to facilitate aeration. The addition of ficoll to the medium to keep protoplasts floating and thereby allowing better growth.

Hanging Drop Method

Kao and his group developed the hanging drop method in 1970 which was subsequently used by others. In this method, a suspension of protoplasts at a density of 104/ml to 105/ml is placed as 50 µl drops in plastic Petri plates, sealed with para film and incubated in an inverted position at 25-30°C under l w light intensity (100-500 lux) or even indark. After cell wall regeneration and the initiation of cell division, fresh medium is added to make cell suspension. The small size of drop helps in providing enough aeration to protoplasts Vasil, 1976.

Agar Plating Method

The method is almost similar to the one developed by Bergmann 1960 to grow callus cells of tobacco and beans and was first employed for the protoplast culture by Nagata and Takabe. The major advantages of this technique are: (i) a large number of protocols can behandled simultaneously, and (ii) plating efficiency can be determined easily. At the same time the major drawback of this technique is that after regeneration of cell wall and induction of cell division, osmotic potential of the medium can not be altered by addition of fresh medium lacking osmotic stabilizer, since the initial medium is semi-solid. However, this problem can be solved by transferring blocks of agar (in which protoplasts are cultured initially) to a fresh medium with lesser concentration of osmoticum or altogether in its absence. This technique has been modified in different ways viz., as feeder technique to support division of cells plated at low density, as micro vessel, and as multiple drop arrays.

Micro Culture Technique

This technique developed by Jones et al. (1960) for culturing isolated cells was used by Vasiland Hilderbrandt 1965 to raise tobacco plants from isolated cells. Micro culture technique has been successfully used for culturing protoplasts of tobacco and Petunia Durand et al., 1973. Adrop of culture medium containing one or more protoplasts is put on a microscopic slide. On either sides of this microscopic slide are kept two cover slips. A third one is put over these two cover slips to shield the protoplasts suspensions.

Multi Drop Array (MDA) Technique

This is a further improvement of hanging drop method and allows screening of large number of hormonal and nutritional factors (up to 4900) using small amount of plant material.

Factors Affecting Protoplasts Culture

Successful growth and regeneration of protoplasts is dependent on a number of factors ranging from the status of the donor plant to the culture conditions.

Plant Species and Varieties

It is a well documented fact that even very small genetic difference leads to varying protoplast responses to culture conditions.

Plant Age and Organ

Plant Age and Organ A second important factor for the successful culture is the age of the donor plant and developmental stage of the donor part used for isolation of the protoplasts. The stages with respect to their responses are germinating embryos, followed by seedling of one week, plantlets, leaves, juvenile tissue from mature plant and mature tissue. Though having the same genetic information protoplasts isolated from these parts behave differently in culture conditions. Plant regeneration has been most successful from leaf protoplasts of her baceous species.

Pre Culture Conditions

Besides the factors mentioned above the behaviour of protoplasts in culture is highly influenced by climatic factors and pre culture conditions. Different culture conditions of seedling yields protoplasts having different responses when cultured.

Pretreatment to the Tissue, Before Isolation of Protoplasts

Pre treatments such as cold treatment, plasmolysis, and hormone treatment to the tissue increases the chances of recovery of viable protoplasts and their plating efficiency.

Nutritional Requirement for Protoplasts Nutritional requirement for the growth of protoplasts differ from the nutritional requirement for tissues and cell culture. Protoplasts leak in cultures as they are devoid of cell walls, and moreover, they have a greater surface area for the diffusion of metabolites than the tissues. Sothe concentration of different metabolites in protoplasts is less than that in the tissues and cells.To compensate for these loses from protoplasts the number of metabolites provided in their culture media is more.

Protoplasts Density

Protoplasts density is very crucial as it influences the plating efficiency and better surviving of protoplasts. When plated at higher density protoplasts compete with one another while at lower density losses of metabolites from protoplasts is more. The later situation of proto plastleakiness can be circumvented by addition of required metabolites to make medium isotonic.Alternatively, one can use nurse callus or feeder layer technique to provide essential organic molecules to protoplast for survival and their subsequent growth.

References

- Callus-culture-meaning-nature-and-significance/42994: biologydiscussion.com, Retrieved 27 May 2018

- Plant-Micropropagation-242579664-1: researchgate.net, Retrieved 12 March 2018

- Cell-suspension-culture-plant-tissue-culture: urbanpro.com, Retrieved 05 July 2018

- Biotech-embryo-culture, plant-biotechnology, In-vitro-culture-techniques, biotechnology: biocyclopedia.com, Retrieved 27 April 2018

- Protoplast-Culture-and-Somatic-Hybridization-233540386: researchgate.net, Retrieved 15 March 2018

Chapter 4

Ploidy in Plants

Ploidy refers to the number of complete sets of chromosomes in a cell. This chapter explores ploidy in plants through a detailed discussion of ploidy, haploid plants, diploid plants, plant life cycles and polyploidy plants.

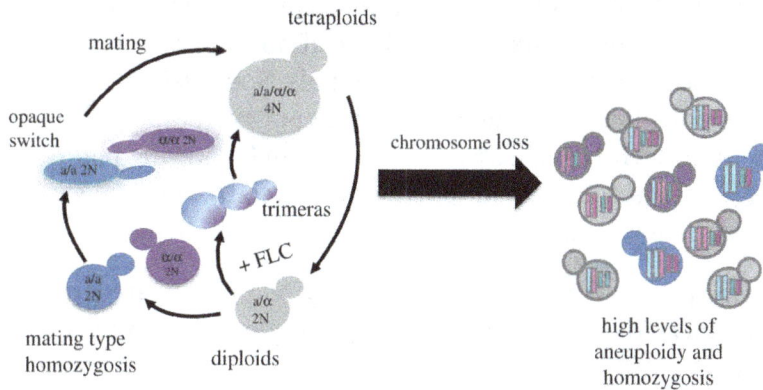

Ploidy in genetics is the number of chromosomes occurring in the nucleus of a cell. In normal somatic (body) cells, the chromosomes exist in pairs. The condition is called diploidy. During meiosis the cell produces gametes, or germ cells, each containing half the normal or somatic number of chromosomes. This condition is called haploidy. When two germ cells (e.g., egg and sperm) unite, the diploid condition is restored.

Polyploidy refers to cells the nuclei of which have three or more times the number of chromosomes found in haploid cells. This condition frequently occurs in plants and may result from chromosome duplication without division of the cytoplasm or from the union of two diploid gametes.

Some cells have an abnormal number of chromosomes that is not a whole multiple of the haploid number. This condition, called aneuploidy, is most often caused by some error resulting in an unequal distribution of chromosomes to the daughter cells. Organisms in which aneuploidy occurs may deviate noticeably from the norm in appearance and behaviour.

Some kinds of trout or salmon have 4 sets (tetraploid). Other examples:

- wheat: 6 sets (hectaploid)

- certain sturgeons: 8 sets (octoploid)

- strawberries: 8 sets (octoploid)

- plumed cockscomb (a plant, Celosia argentea): 12 sets (dodecaploid)

- Bryophytes: body has one set; sporophyte has two sets.

Sex cells (gametes) are almost always haploid. If the reproductive stage (adult) is polyploid, then the gametes will have half the ploidy number of chromosomes.

There are some variations of ploidy which are not discussed here. Some plant species with certain types of polyploidy do not use sexual reproduction, but survive with asexual methods.

Polyploidy

Polyploidy occurs in cells and organisms when there are more than two paired (homologous) sets of chromosomes.

Most organisms are normally diploid, meaning they have two sets of chromosomes — one set inherited from each parent. Polyploidy may occur due to abnormal cell division. It is most commonly found in plants, but it does sometimes happen in animals. Some estimates suggest that 30-80% of living plant species are polyploid, and many lineages show evidence of ancient polyploidy (paleopolyploidy) in their genomes. Huge increases in angiosperm (flowering plants) diversity have coincided with the timing of ancient genome duplications shared by many species. 15% of angiosperm and 31% of fern speciation events are accompanied by ploidy increase.

Polyploid plants arise spontaneously in nature. Many polyploids are fitter than their parental species, and may display novel variation or morphologies that contribute to speciation and eco-niche exploitation.

Polyploidy may occur in one generation, and is an exception to the principle that evolution occurs gradually. There may, however, be many genetic changes in the species after polyploidy has taken place.

Gametes

The gametes of polyploids are unusual, because they may carry several sets of chromosomes. For example, common wheat is a polyploid with six sets of chromosomes, two sets coming originally from each of three different species. So there are six sets of chromosomes in most cells, and three sets of chromosomes in the gametes.

Haploid Plants

A haploid is a cell or organism that has a single set of chromosomes that are not paired. The haploid gamete is normally produced during plant cell division. During fertilization, these cells normally merge with other similar haploid cells. A haploid cell only has half the number of chromosomes as are present in diploids.

There are two approaches for the production of haploid plants. The two approaches are: (1) In Vivo Approach and (2) In Vitro Approach.

Haploid plants are characterized by possessing only a single set of chromosomes (gametophytic number of chromosomes i.e. n) in the sporophyte. This is in contrast to diploids which contain two sets (2n) of chromosomes. Haploid plants are of great significance for the production of homozygous lines (homozygous plants) and for the improvement of plants in plant breeding programmes.

Grouping of Haploids

Haploids may be divided into two broad categories:

Monoploids (Monohapioids)

These are the haploids that possess half the number of chromosomes from a diploid species e.g. maize, barley.

Polyhaploids

The haploids possessing half the number of chromosomes from a polyploid species are regarded as polyhaploids e.g. wheat, potato. It may be noted that when the term haploid is generally used it applies to any plant originating from a sporophyte (2n) and containing half the number (n) of chromosomes.

In Vivo and in Vitro Approaches

The importance of haploids in the field of plant breeding and genetics was realised long ago. Their practical application, however, has been restricted due to very a low frequency (< 0.001%) of their formation in nature.

The process of apomixis or parthenogenesis (development of embryo from an unfertilized egg) is responsible for the spontaneous natural production of haploids. Many attempts were made, both by in vivo and in vitro methods to develop haploids. The success was much higher by in vitro techniques.

In Vivo Techniques for Haploid Production

There are several methods to induce haploid production in vivo.

Some of them are Listed Below

Androgenesis

Development of an egg cell containing male nucleus to a haploid is referred to as androgenesis. For a successful in vivo androgenesis, the egg nucleus has to be inactivated or eliminated before fertilization.

Gynogenesis

An unfertilized egg can be manipulated (by delayed pollination) to develop into a haploid plant.

Distant Hybridization

Hybrids can be produced by elimination of one of the parental genomes as a result of distant (interspecific or inter-generic crosses) hybridization.

Irradiation Effects

Ultra violet rays or X-rays may be used to induce chromosomal breakage and their subsequent elimination to produce haploids.

Chemical Treatment

Certain chemicals (e.g., chloramphenicol, colchicine, nitrous oxide, maleic hydrazide) can induce chromosomal elimination in somatic cells which may result in haploids.

In vitro techniques for haploid production

In the plant biotechnology programmes, haploid production is achieved by two methods.

Androgenesis

Haploid production occurs through anther or pollen culture, and they are referred to as androgenic haploids.

Gynogenesis

Ovary or ovule culture that results in the production of haploids, known as gynogenic haploids.

Androgenesis

In androgenesis, the male gametophyte (microspore or immature pollen) produces haploid plant. The basic principle is to stop the development of pollen cell into a gamete (sex cell) and force it to develop into a haploid plant. There are two approaches in androgenesis— anther culture and pollen (microspore) culture. Young plants, grown under optimal conditions of light, temperature and humidity, are suitable for androgenesis.

Anther Culture

The selected flower buds of young plants are surface-sterilized and anthers removed along with their filaments. The anthers are excised under aseptic conditions, and crushed in 1% acetocarmine to test the stage of pollen development.

If they are at the correct stage, each anther is gently separated (from the filament) and the intact anthers are inoculated on a nutrient medium. Injured anthers should not be used in cultures as they result in callusing of anther wall tissue.

The anther cultures are maintained in alternating periods of light (12-18 hr.) and darkness (6-12 hrs.) at 28°C. As the anthers proliferate, they produce callus which later forms an embryo and then a haploid plant.

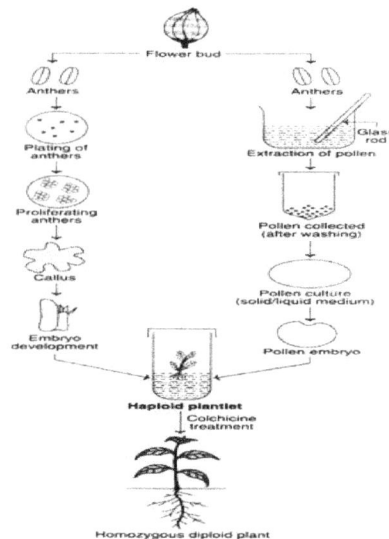

Pollen (Microspore) Culture

Haploid plants can be produced from immature pollen or microspores (male gametophytic cells). The pollen can be extracted by pressing and squeezing the anthers with a glass rod against the sides of a beaker. The pollen suspension is filtered to remove anther tissue debris.

Viable and large pollen (smaller pollen do not regenerate) are concentrated by filtration, washed and collected. These pollen are cultured on a solid or liquid medium. The callus/embryo formed is transferred to a suitable medium to finally produce a haploid plant, and then a diploid plant (on colchicine treatment).

Comparison between Anther and Pollen Cultures

Anther culture is easy, quick and practicable. Anther walls act as conditioning factors and promote culture growth. Thus, anther cultures are reasonably efficient for haploid production. The major limitation is that the plants not only originate from pollen but also from other parts of anther. This results in the population of plants at different ploidy levels (diploids, aneuploids). The disadvantages associated with anther culture can be overcome by pollen culture.

Many workers prefer pollen culture, even though the degree of success is low, as it offers the following advantages:

- Undesirable effects of anther wall and associated tissues can be avoided.

- Androgenesis, starting from a single cell, can be better regulated.

- Isolated microspores (pollen) are ideal for various genetic manipulations (transformation, mutagenesis).

- The yield of haploid plants is relatively higher.

Development of Androgenic Haploids

The process of in vitro androgenesis for the ultimate production of haploid plants is depicted in figure below.

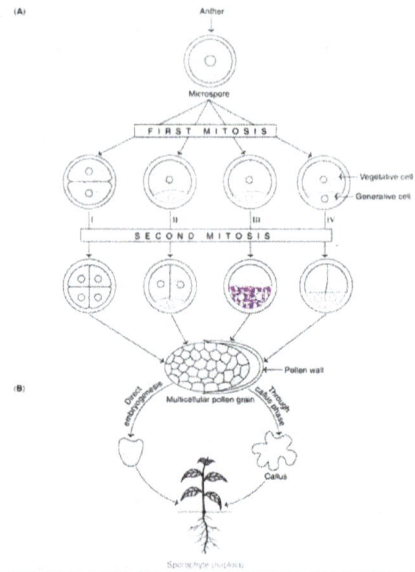

The cultured microspores mainly follow four distinct pathways during the initial stages of in vitro androgenesis.

Pathway I

The uninucleate microspore undergoes equal division to form two daughter cells of equal size e.g. Datura innoxia.

Pathway II

In certain plants, the microspore divides unequally to give bigger vegetative cell and a smaller generative cell. It is the vegetative cell that undergoes further divisions to form callus or embryo. The generative cell, on the other hand, degenerates after one or two divisions—e.g., Nicotiana tabacum, Capsicum annuum.

Pathway III

In this case, the microspore undergoes unequal division. The embryos are formed from the generative cell while the vegetative cell does not divide at all or undergoes limited number of divisions e.g. HyoScyamus niger.

Pathway IV

The microspore divides unequally as in pathways I and II. However, in this case, both vegetative and generative cells can further divide and contribute to the development of haploid plant e.g. Datura metel, Atropa belladonna.

At the initial stages, the microspore may follow any one of the four pathways described above. As the cells divide, the pollen grain becomes multicellular and burst open. This multicellular mass may form a callus which later differentiates into a plant (through callus phase). Alternately, the multicellular mass may produce the plant through direct embryogenesis.

Factors Affecting Androgenesis

A good knowledge of the various factors that influence androgenesis will help to improve the production of androgenic haploids.

Genotype of Donar Plants

The success of anther or pollen culture largely depends on the genotype of the donor plant. It is therefore important to select only highly responsive genotypes. Some workers choose a breeding approach for improvement of genotype before they are used in androgenesis.

Stage of Microspore or Pollen

The selection of anthers at an ideal stage of microspore development is very critical for haploid production. In general, microspores ranging from tetrad to bi-nucleate stages are more responsive. Anthers at a very young stage (with microspore mother cells or tetrads) and late stage (with bi-nucleate microspores) are usually not suitable for androgenesis. However, for maximum production of androgenic haploids, the suitable stage of microspore development is dependent on the plant species, and has to be carefully selected.

Physiological Status of a Donar Plant

The plants grown under best natural environmental conditions (light, temperature, nutrition, CO_2 etc.) with good anthers and healthy microspores are most suitable as donor plants. Flowers obtained from young plants, at the beginning of the flowering season are highly responsive. The use of pesticides should be avoided at least 3-4 weeks preceding sampling.

Pretreatment of Anthers

The basic principle of native androgenesis is to stop the conversion of pollen cell into a gamete, and force its development into a plant. This is in fact an abnormal pathway induced to achieve in vitro androgenesis. Appropriate treatment of anthers is required for good success of haploid production.

Treatment methods are variable and largely depend on the donor plant species.

Chemical Treatment

Certain chemicals are known to induce parthenogenesis e.g. 2-chloroethylphosphonic acid (ethrel). When plants are treated with ethreal, multinucleated pollens are produced. These pollens when cultured may form embryos.

Temperature Influence

In general, when the buds are treated with cold temperatures (3-6°C) for about 3 days, induction occurs to yield pollen embryos in some plants e.g. Datura, Nicotiana. Further, induction of androgenesis is better if anthers are stored at low temperature, prior to culture e.g. maize, rye. There are also reports that pretreatment of anthers of certain plants at higher temperatures (35°C) stimulates androgenesis e.g. some species of Brassica and Capsicum.

Effect of Light

In general, the production of haploids is better in light. There are however, certain plants which can grow well in both light and dark. Isolated pollen (not the anther) appears to be sensitive to light. Thus, low intensity of light promotes development of embryos in pollen cultures e.g. tobacco.

Effect of culture medium

The success of another culture and androgenesis is also dependent on the composition of the medium. There is, however, no single medium suitable for anther cultures of all plant species. The commonly used media for anther cultures are MS, White's, Nitsch and Nitsch, N6 and B5. These media in fact are the same as used in plant cell and tissue cultures. In recent years, some workers have developed specially designed media for anther cultures of cereals.

Sucrose, nitrate, ammonium salts, amino acids and minerals are essential for androgenesis. In some species, growth regulators — auxin and/or cytokinin are required for optimal growth. In certain plant species, addition of glutathione and ascorbic acid promotes androgenesis. When the anther culture medium is supplemented with activated charcoal, enhanced androgenesis is observed. It is believed that the activated charcoal removes the inhibitors from the medium and facilitates haploid formation.

Gynogenesis

Haploid plants can be developed from ovary or ovule cultures. It is possible to trigger female gametophytes (megaspores) of angiosperms to develop into a sporophyte. The plants so produced are referred to as gynogenic haploids.

Gynogenic haploids were first developed by San Noem (1976) from the ovary cultures of Hordeum vulgare. This technique was later applied for raising haploid plants of rice, wheat, maize, sunflower, sugar beet and tobacco.

In vitro culture of un-pollinated ovaries (or ovules) is usually employed when the anther cultures give .unsatisfactory results for the production of haploid plants. The procedure for gynogenic haploid production is briefly described.

The flower buds are excised 24-48 hr. prior to anthesis from un-pollinated ovaries. After removal of calyx, corolla and stamens, the ovaries are subjected to surface sterilization. The ovary, with a cut end at the distal part of pedicel, is inserted in the solid culture medium.

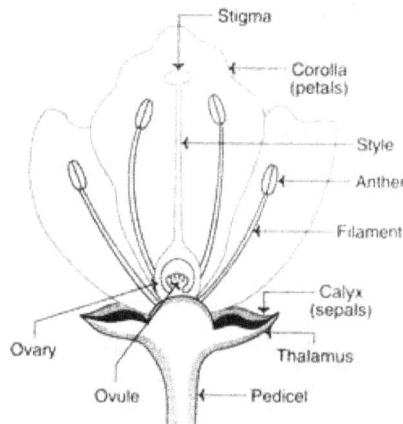

Whenever a liquid medium is used, the ovaries are placed on a filter paper or allowed to float over the medium with pedicel inserted through filter paper. The commonly used media are MS, White's, N6 and Nitsch, supplemented growth factors. Production of gynogenic haploids is particularly useful in plants with male sterile genotype. For such plant species, this technique is superior to another culture technique.

Limitations of Gynogenesis

In practice, production of haploid plants by ovary/ ovule cultures is not used as frequently as anther/ pollen cultures in crop improvement programmes.

The major limitations of gynogenesis are listed

1. The dissection of unfertilized ovaries and ovules is rather difficult.

2. The presence of only one ovary per flower is another disadvantage. In contrast, there are a large number of microspores in one another.

However, the future of gynogenesis may be more promising with improved and refined methods.

Identification of Haploids

Two approaches based on morphology and genetics are commonly used to detect or identify haploids.

*   Morphological Approach:

    The vegetative and floral parts and the cell sizes of haploid plants are relatively reduced when compared to diploid plants. By this way haploids can be detected in a population of diploids. Morphological approach, however, is not as effective as genetic approach.

*   Genetic Approach:

    Genetic markers are widely used for the specific identification of haploids. Several markers are in use.

    ◦   'a1' marker for brown coloured aleurone.

- ◦ 'A' marker for purple colour.

- ◦ 'Lg' marker for ligule less character.

The above markers have been used for the development of haploids of maize. It may be noted that for the detection of androgenic haploids, the dominant gene marker should be present in the female plant.

Diploidizatioim of Haploid Plants (Production of Homozygous Plants)

Haploid plants are obtained either by androgenesis or gynogenesis. These plants may grow up to a flowering stage, but viable gametes cannot be formed due to lack of one set of homologous chromosomes. Consequently, there is no seed formation.

Haploids can be diploidized (by duplication of chromosomes) to produce homozygous plants. There are mainly two approaches for diploidization— colchicine treatment and endomitosis.

Colchicine Treatment

Colchicine is very widely used for diploidization of homologous chromosomes. It acts as an inhibitor of spindle formation during mitosis and induces chromosome duplication. There are many ways of colchicine treatment to achieve diploidization for production of homozygous plants.

1. When the plants are mature, colchicine in the form of a paste is applied to the axils of leaves. Now, the main axis is decapitated. This stimulates the axillary buds to grow into diploid and fertile branches.

2. The young plantlets are directly treated with colchicine solution, washed thoroughly and replanted. This results in homozygous plants.

3. The axillary buds can be repeatedly treated with colchicine cotton wool for about 2-3 weeks.

Endomitosis

Endomitosis is the phenomenon of doubling the number of chromosomes without division of the nucleus. The haploid cells, in general, are unstable in culture with a tendency to undergo endomitosis. This property of haploid cells is exploited for diploidization to produce homozygous plants.

The procedure involves growing a small segment of haploid plant stem in a suitable medium supplemented with growth regulators (auxin and cytokinin). This induces callus formation followed by differentiation. During the growth of callus, chromosomal doubling occurs by endomitosis. This results in the production of diploid homozygous cells and ultimately plants.

Diploid Plants

Diploid describes a cell or nucleus which contains two copies of genetic material, or a complete set of chromosomes, paired with their homologs (chromosome carrying the same information from the other parent). By maintaining two copies of the genetic code, diploid organisms obtain an advantage by having greater genetic variation within their population, as each individual can express two alleles for each gene. Other organisms cycle between diploid and haploid lifecycles.

Example of Diploid

Lifecycle of a Fern

Unlike humans and other mammals, ferns have an entire multi-celled stage of their lifecycle which is not diploid. Look at the diagram below. During the sporophyte phase, the plant is diploid. This diploid plant creates spores through meiosis, which are now haploid. The haploid cells are released into the air and travel to a new area.

Once established, the haploid spore grows into a gametophyte. The gametophyte is an entire haploid organism, separate from the first plant. This small plant has special tissues which create gametes in the form of sperm and eggs. These haploid cells find each other and fertilize one another, creating diploid zygotes. These zygotes then grow into full sporophytes, and the cycle starts over. Where humans and many familiar animals spend the entirety of their lives as diploid organisms, many species such as ferns and insects are not that way.

Polyploid Plants

Polyploids are organisms with multiple sets of chromosomes in excess of the diploid number. Polyploidy is common in nature and provides a major mechanism for adaptation and speciation. Approximately 50-70% of angiosperms, which include many crop plants, have undergone polyploidy during their evolutionary process. Flowering plants form polyploids at a significantly high frequency of 1 in every 100,000 plants. Many studies have been carried out to understand the nature of polyploidism. To understand polyploidy, a few basic notations need be defined. The basic complete set of chromosomes is designated by "x" while the total number of chromosomes in a somatic cell is designated "2n". The total number of chromosomes in a somatic cell is twice the haploid number (n) in the gametes. The ploidy of some of the major crops in the world is represented in Table below.

Classification of Polyploids

Polyploids may be classified based on their chromosomal composition into either euploids or aneuploids. Euploids constitute the majority of polyploids.

Euploidy

Euploids are polyploids with multiples of the complete set of chromosomes specific to a species. Depending on the composition of the genome, euploids can be further classified into either autopolyploids or allopolyploids. Tetraploidy is the most common class of euploids.

A list of major crops and their ploidy

Common name	Ploidy	Name	Propagation
Maize	2x=20	Diploid	Outcrossing
Wheat	6x=42	Hexaploid	Outcrossing
Rice	2x=24	Diploid	Selfing
Potatoes	4x=48	Tetraploid	Outcrossing; Vegetative
Soybeans	2x=40	Diploid	Selfing
Barley	2x=14	Diploid	Selfing
Tomatoes	2x=24	Diploid	Selfing
Bananas	3x=33	Triploid	Vegetative
Watermelon	2x=22	Diploid	Outcrossing
Sugarcane	8x=80	Octoploid	Outcrossing; vegetative
Sugar beet	2x=18	Diploid	Outcrossing
Cassava	2x=36	Diploid	Outcrossing; Vegetative

Autopolyploidy

Autopolyploids are also referred to as autoploids. They contain multiple copies of the basic set (x) of chromosomes of the same genome. Autoploids occur in nature through union of unreduced gametes and at times can be artificially induced.

Natural autoploids include tetraploid crops such as alfafa, peanut, potato and coffee and triploid bananas. They occur spontaneously through the process of chromosome doubling. Chromosome doubling in autoploids has varying effect based on the species. Spontaneous chromosome doubling in ornamentals and forage grasses has led to increased vigour. For instance, ornamentals such as tulip and hyacinth, and forage grasses such as ryegrasses have yielded superior varieties following spontaneous chromosome doubling. Due to the observed advantages in nature, breeders have harnessed the process of chromosome doubling in vitro through induced polyploidy to produce superior crops. For example, induced autotetraploids in the watermelon crop are used for the production of seedless triploid hybrids fruits. Such polyploids are induced through the treatment of diploids with mitotic inhibitors such as dinitroaniles and colchicine. To determine the ploidy status of induced polyploids, several approaches may be used. These include, chloroplast count in guard cells, morphological features such as leaf, flower or pollen size (gigas effect) and flow cytometry.

Allopolyploidy

Allopolyploids are also called alloploids. They are a combination of genomes from different species. They result from hybridization of two or more genomes followed by chromosome doubling or by the fusion of unreduced gametes between species. This process is key in the process of speciation for angiosperms and ferns and occurs often in nature. Economically important natural alloploid crops include strawberry, wheat, oat, upland cotton, oilseed rape, blueberry and mustard. To differentiate between the sources of the genomes in an alloploid, each genome is designated by a different letter. For example, the origin of the cultivated mustards (Brassica spp) has been well explained by Nagaharu in the triangle of U with each species represented by a distinct letter.

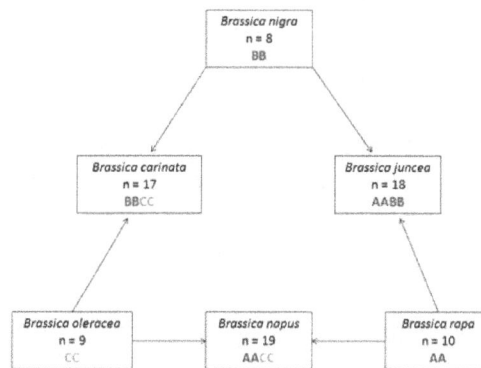

The hybridized genomes differ in their degree of homology with some being able to pair during mitosis and/or meiosis while others not. When only segments of the chromosomes of the combining genomes differ, the phenomenon is called segmental alloploidy. These chromosomes are similar but not homologous and are called homeologous chromosomes. Such chromosomes indicate ancestral homology. Induced alloploidy is not common. However, it has been used in some Genus such as Cucumis to elucidate the molecular mechanisms involved in diploidization (tendency of polyploids to act as diploids). In this study, an allotetraploid was induced by hybridization between Cucumis sativus and Cucumis hystrix followed by chromosome doubling. Cytogenetic studies were carried out in the following generations to establish the molecular mechanisms involved stabilization of newly formed allopolyploids which include neo-functionalization and sub-functionalization.

Aneuploidy

Aneuploids are polyploids that contain either an addition or subtraction of one or more specific chromosome(s) to the total number of chromosomes that usually make up the ploidy of a species. Aneuploids result from the formation of univalents and multivalents during meiosis of euploids. For example, several studies have found that 30-40% of progeny derived from autotetraploid maize are aneuploids. With no mechanism of dividing univalents equally among daughter cells during anaphase I, some cells inherit more genetic material than others. Similarly, multivalents such as homologous chromosomes may fail to separate during meiosis leading to unequal migration of chromosomes to opposite poles. This mechanism is called non-disjunction. These meiotic aberrances result in plants with reduced vigor. Aneuploids are classified according to the number of chromosomes gained or lost as shown in Table below.

Classification of Aneuploids

Term	Chromosome number
Monosomy	2n-1
Nullisomy	2n-2
Trisomy	2n+2
Tetrasomy	2n+2
Pentasomy	2n+3

Mechanisms of Polyploidy Formation

Several cytological mechanisms are known to spontaneously induce polyploidy in plants. One such route involves non-reduction of gametes during meiosis a process called meiotic nuclear restitution. The formed gametes (2n) contain the somatic nuclear condition of cells. Meiotic aberrations related to spindle formation, spindle function and cytokinesis have been implicated in this process. The subsequent union of reduced and non-reduced gametes leads to the formation of polyploids. For example, autotetraploids may be formed in a diploid population through the union of two unreduced 2n gametes as was found in the F1 progenies of open-pollinated diploid apples. Similarly, spontaneous allotetraploids were formed in 90% of F2 progenies of interspecific crosses between Digitalis ambigua and Digitalis purpurea, which are common ornamental plants. Another example is the formation of autohexaploid Beta vulgaris sugar beet and alfalfa from cultivated autotetraploid varieties apparently from the union of reduced (2x) and unreduced (4x) gametes.

Another major route for polyploid formation is through somatic doubling of chromosomes during mitosis. In nature, the formation of polyploids as a result of mitotic aberrations has been reported in the meristematic tissue of several plant species including tomato and in non-meristematic tissues of plants such as bean. Artificial inducement of polyploids through the inhibition of mitosis is routine in plant breeding. High temperatures above 40°C have been used to induce tetraploid and octoploid corn seedlings albeit with low success of 1.8% and 0.8% respectively. Currently, chemical mitotic inhibitory agents such as colchicine or dinitroanilines are used to induce polyploidy in crop plants. A typical example is the production of tetraploid watermelon plants for the production of seedless triploid watermelon.

In addition, an uncommon mechanism of polyploid formation involves polyspermy where one egg is fertilized by several male nucleuses as commonly observed in orchids. The major pathways involved in polyploidy formation are represented in figure below.

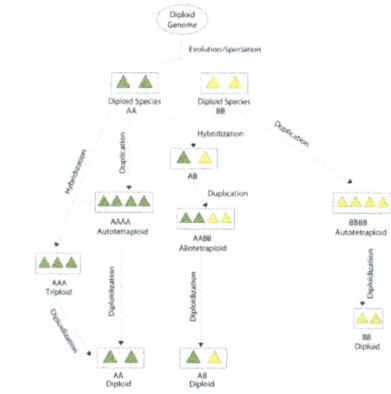

Alterations Associated with Polyploidy

Several changes in the plant accompany spontaneous or induced polyploidy. These may be changes in genetic composition, physiological mechanisms, structural composition and vigor. Some of these changes create the platform for the commercial exploitation of polyploids. Genetic changes following genome duplication involve the rapid loss of chromosomal segments in a process called diploidization. Diploidization describes the process by which a polyploid genome become more 'diploid-like' in character. It is necessary to eliminate duplicated genes in a newly formed polyploid to avoid gene silencing as well as to stabilize fertility. Duplicated genes that are retained often undergo subfunctionalization (complementing genes) and neofunctionalization (genes with novel functions). Diploidization has been described for many genus including Nicotina and Cucumis.

The increase in nuclear ploidy affects the structural and anatomical characteristics of the plant. In general, polyploidy results in increased leaf and flower size, stomatal density, cell size and chloroplast count. These phenomena are collectively referred to as the gigas effect.

Physiological changes are also known to accompany genome duplication. These mainly result from change of metabolism resulting in a general increase in secondary metabolites. This property has found application in the breeding of medicinal herbs in the production of pharmaceuticals. Hybrid vigor resulting from interspecific crosses in allopolyploids is one of the most exploited advantages of polyploid in plant breeding.

Figure: A comparison between the leaf and flower of a (A) diploid and (B) induced tetraploid watermelon illustrating the gigas effect.

Implications of Ploidy in Plant Breeding

Heterosis in Allopolyploids

Heterosis or hybrid vigor is the difference between the hybrid and the mean of the two parents and is characterized by increased vigor and superior qualitative or quantitative traits. Over the last several decades, breeders have increased the world food production by utilizing the concept of heterosis in hybrid cultivars. For example, following the introduction of hybrid corn diploid in the 1920's, there was a six fold increase in corn production between then and 1990 in the U.S. However, unlike diploids which may lose heterosis with each consecutive generation due to segregation, alloploidy and autoploidy imposes pairing of homologous chromosomes, thus preventing intergenomic recombination. This concept is called preferential or selective pairing and is the tendency for a doubled set of chromosomes to pair independently of the doubled set of chromosomes of the other species. In this way, heterozygosity is maintained throughout generations. Generally, the parents used in hybrid formation should be within subspecies or between subspecies. An example of a man-made interspecies allopolyploid hybrid is triticale. It is derived from crossing two cereals, hexaploid bread wheat (T. aestivum) and rye (Secale cereale). Triticale was developed to combine good qualities of wheat including high yield and grain quality with the hardiness (disease and stress tolerance) of rye.

The process of hybrid formation for polyploids is not without setbacks. Many interspecific hybrids have low fertility and viability due to hybrid incompatibilities. Hybrid incompatibility results from genes that are functionally diverged in the respective hybrid forming species. This may lead to silencing of protein encoding genes and has been reported in interspecific hybrids of Arabidopsis. To increase the heterosis, fertility and viability of interspecific hybrids, several factors should be considered. The parents used should be of diverse genetic background and preferably heterozygous.

Inbreeding in Polyploids

Self pollination is an important method for attaining homozygosity in breeding. Through this process, it is possible to fix desired alleles in the background of a crop. In general, it takes approximately 3.80 more generations for an autotetraploid to reach the same level of homozygosity as the diploid. Fixing a trait controlled by a single gene in an autotetraploid, would require four identical alleles to achieve homozygosity. For example, in a segregating, tetraploid F2 population the proportion of the homozygous loci would be 1:18.

Effect of Polyploidy on Inheritance and Population Genetics

An immediate consequence of polyploidy is the change in gametic and filial frequencies. This is because polyploids have multiple alleles associated with a single locus. For example, a hexaploid has six alleles per locus while a tetraploid has four. The genetics of polyploids is often complicated by multi-allelism at loci thus altering segregation ratios and inheritance patterns expected in diploids. Provided a polyploid species behaves like a diploid at meiosis through normal bivalent pairing (disomic inheritance), such as in wheat or tobacco, normal biometric analysis of inheritance apply. However, several autotetraploid crop plants including potatoes, coffee and lucerne and some forage grasses have tetrasomic inheritance. With this knowledge, it is necessary to make accommodations in population structure and breeding strategy to account for differences in gamete structure. For example, breeding schemes that maximize heterozygosity are frequently used for the autotetraploid alfalfa in an attempt to utilize multi-allelic interactions. Altered genotypic ratios are apparent in polyploids when compared with diploids. For example an arbitrary locus with B (dominant) and b (recessive) alleles, following selfing, an autotetraploid (BBbb) would produce 5 possible genotypes while a diploid (Bb) would generate 3 possible genotypes. Distinguishing a quadruplex (BBBB) from a triplex (BBBb) in the segregating population using a progeny test presents difficulty in breeding because both would breed true to the dominant allele. An extra generation would be required to identify the triplex by observing the formation of duplex plants.

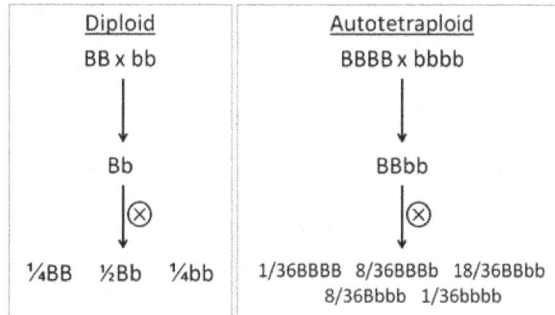

Figure: Genotypes and ratios formed following selfing of a heterozygous diploid and autotetraploid.

Effect of Polyploidy on Sterility

Since autoploids contain more than two homologous chromosomes, meiosis results in the formation of univalents and multivalent, unlike in diploids where bivalents are usually formed. For instance during meiosis, autotetraploids may form bivalents, quadrivalents and univalents. The ratio of these gametes following meiosis determines the fertility of a polyploid individual. Univalents and trivalents result in non-functional sterile gametes and are the most common in triploids, making them sterile.

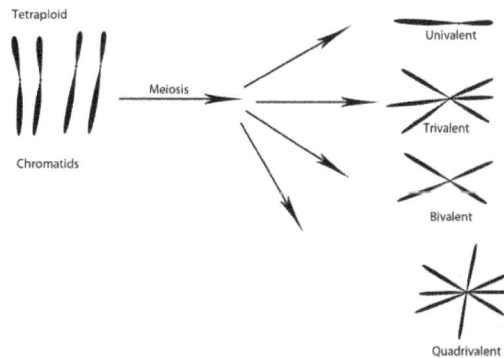

Figure: Gamete formation in autotetraploids. Trivalents and univalents result in s terile gametes while bivalents and quadrivalents result in fertile gametes.

Rigorous and effective selection strategies for fertile autoploids are practiced in the development of inbred lines. Breeders rogue out autoploids with low seed set as well as those with morphological abnormalities. Sterile alloploids arise from the pairing of homeologous chromosomes from separate genomes during meiosis instead of homologous chromosome. This results in non-functional gametes. A viable allopolyploid requires a diploid-like meiosis behavior to establish disomic inheritance and full fertility. Fertility problems in allopolyploids also occur when crossing crops of different ploidy levels as a result of formation of multivalents. To improve fertility, breeders use the parent with the lowest chromosome number as the female parent so as to increase seed set.

Common Applications of Ploidy in Crop Plants

Mutation Breeding

High frequencies of chromosome mutations are desirable in modern breeding techniques, such as tilling, as they provide new sources of variation. The multiallelic nature of loci in polyploids has many advantages that are useful in breeding. The masking of deleterious alleles, that may arise from induced mutation, by their dominant forms cushions polyploids from lethal conditions often associated with inbred diploid crops. This concept has been instrumental in the evolution of polyploids during bottlenecks where there is enforced inbreeding. Mutation breeding exploits the concept of gene redundancy and mutation tolerance in polyploid crop improvement in two ways. First, polyploids are able to tolerate deleterious allele modifications post-mutation, and secondly, they have increased mutation frequency because of their large genomes resulting from duplicated condition of their genes. The high mutation frequencies observed with polyploids may be exploited when trying to induce mutations in diploid cultivars that do not produce enough genetic variation after a mutagenic treatment. This approach has been used in mutation breeding of Achimenes sp. (nut orchids) by first forming autotetraploids through colchicine treatment followed by the application of fast neutrons and X-rays. In this study, the autotetraploids were found to have 20-40 times higher mutation frequency than the corresponding diploid cultivar due to the large genome.

Seedless Fruits

The seedless trait of triploids has been desirable especially in fruits. Commercial use of triploid fruits can be found in crops such as watermelons and are produced artificially by first developing tetraploids which are then crossed with diploid watermelon. In order to set fruit, the triploid watermelon is crossed with a desirable diploid pollen donor.

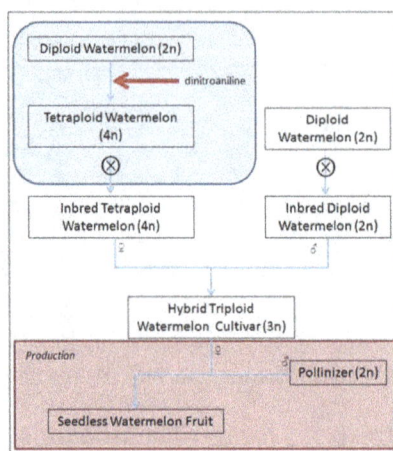

Bridge Crossing

Another breeding strategy that utilizes the reproductive superiority of polyploids is bridge crossing. When sexual incompatibilities between two species are due to ploidy levels, transitional crosses can be carried out followed by chromosome doubling to produce fertile bridge hybrids. This method has been used to breed for superior tall fescue grass (F. arundinacea) from Italian ryegrass (2n=2x=14) and tall fescue (2n=6x=42) by using meadow grass (Fescue pratensis) as a bridge species. The same principle has been applied in fixing heterozygosity in hybrids by doubling the chromosomes in the superior progeny.

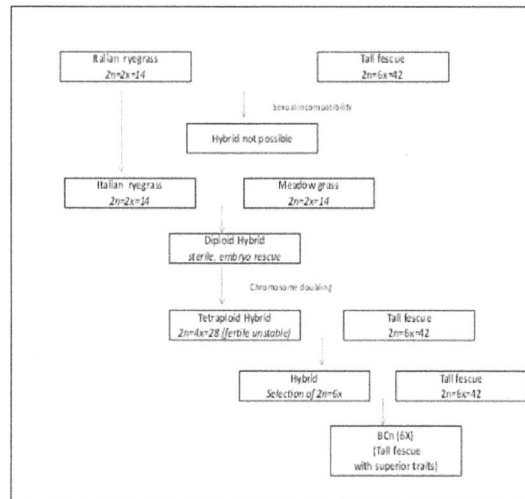

Figure: The development of superior tall fescue grass through bridge crossing and induced tetraploidy.

Ornamental and Forage Breeding

One of the immediate and obvious consequences of polyploidy in plants is an increase in cell size which in turn leads to enlarged plant organs, a phenomenon termed gigas effect. For example, the volume of tetraploid cells usually is about twice that of their diploid progenitors. The increase in cell volume however is mainly attributed to increased water and not biomass. Therefore, its application is limited for breeding agronomically important crops such as cereals. Although chromosome doubling may result in significantly larger seeds and increased seed-protein content in cereal crops, this advantage is offset by low seed set. In contrast, the gigas effect has been explored in tree, ornamental, forage crop and fruit breeding. For example, through induced polyploidy, breeders have developed Bouschet tetraploid grapes with more yield and juice content than the diploid progenitor Alicante. Ornamental crops such as snapdragons and marigolds have been bred through chromosome doubling to improve the quality and size of their blossoms. A strong inverse correlation between DNA content and development rates in plants has been reported by several authors. It has been attributed to lower auxin levels, reduced surface to volume ratio and altered nuclear surface to cell volume ratio. The slower growth rate of polyploids allows them to flower later and for a longer period of time than their diploid progenitors. This quality may be of interest especially in ornamental breeding.

Production of Apomictic Crops

Apomixis provides another avenue for use of polyploids in breeding. Apomixis provides an avenue for the production of seeds asexually through parthenogenesis. Most apomictic plants are polyploid but most polyploid plants are not apomictic. In plants capable of both sexual and asexual reproduction, polyploidy promotes the latter. Obligate apomicts are the most desired of hybrids but little gain has been realized towards their development. However, it has been suggested that obligate apomicts may be induced through development of very high ploidy plants. An example of an obligate apomict achieved at high ploidy level is the octoploid of the grass, Themeda triandra.

Disease Resistance Through Aneuploidy

Aneuploidy has been applied in breeding to develop disease resistant plants through the addition of an extra chromosome into the progeny genome. An example is the transfer of leaf rust resistance to Tricum aestivum from Aegilops umbellulata through backcrossing. In addition, other breeding strategies utilizing aneuploidy have been explored including chromosome deletion, chromosome substitution and supernumerary chromosomes.

Industrial Applications of Polyploidy

Chromosome doubling is reported to have an apparent effect on many physiological properties of a plant. The most discernable of these has been the increase in secondary as well as primary metabolism. The resulting increase in secondary metabolites, in some cases by 100%, after chromosome doubling has been widely exploited in the breeding of narcotic plants such as Cannabis, Datura and Atropa. In vitro secondary metabolite production systems that exploit polyploidism have also been developed. The production of the antimalarial sesquiterpene artemisinin has been enhanced six fold by inducing tetraploids of the wild diploid Artemisia annua L. (clone YUT16). In addition, commercial synthesis of sex hormones and corticosteroids has been improved significantly by artificial induction of tetraploids from diploid Dioscorea zingiberensis, native to China. Attempts have been made to improve the production of pyrethrin, a botanical insecticide, by chromosome doubling of Chrysanthemum cinerariifolium. Other plants whose production of terpenes has increased following artificial chromosome doubling include Carum cari, Ocimum kilmandscharicum and Mentha arvensis. The enhanced production of secondary metabolites such as alkaloids and terpenes in polyploids may concurrently offer resistance to pests and pathogens. Experiments with diploid Glycine tabacina, a forage legume, and its tetraploid forms to measure resistance to leaf rust, Phakopsora pachyrhizi, established that 12% of the tetraploid plants were resistant compared to 14% of the diploid plants. Similar results were observed while comparing resistance to insects and the clover eel nematode between Trifolium pratense (red clover) tetraploids and diploids.

Plant Life Cycles

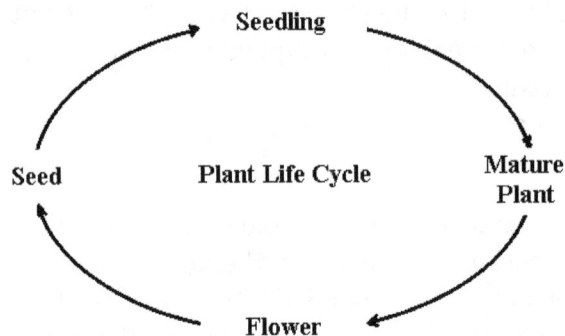

The complete series of events from zygote formation to gamete production constitutes the plant life cycle. des all the processes involved from with the growth, development, and reproduction of plant species. There are three different plant life cycles: haploid (1n), diploid (2n), and the more common haploid-diploid (1n-2n). A haploid organism consists of a multicellular structure of cells that contain only one set of chromosomes, whereas, a diploid organism's multicellular stage contains two sets of chromosomes. Of the haplid organisms, there are some fungi that go through a dikaryotic phase and some plants and fungi (note: fungi are now placed in a separate kingdom from plants, the Fungi kingdom) that lack a dikaryotic phase. Since we are concentrating on the life histories of plants, we will focus on plants that lack the dikaryotic phase. In a 1n-2n life cycle, there are two multicellular generations, a haploid (1n gametophyte) alternating with a diploid (2n sporophyte) stage. Moreover, there is an alternation of either similar or dissimilar generations. In the first case, the 1n gametophyte is morphologically identical to the 2n sporophyte, except for the reproductive structures, while in the latter case, the two generations are morphologically different. For the alternation of dissimilar generations, there are three possible relationships between the gametophyte and the sporophyte: 1. both are totally independent of each other at maturity (each being free-living), 2. the sporophyte is dominant, or 3. the gametophyte is dominant.

Moreover, there are also variations within each of these life histories. Asexual, or vegetative, reproduction may also occur within all three life cycles. Sexual production is characterized by syngamy (fertilization) which involves the fusion of 1n gametes to form a zygote and is followed by meiosis in which a 2n structure divides into four different 1n cells. Asexual reproduction, on the other hand, does not involve these two processes. As for the evolutionary history of these life cycles, haploid alga were predominant prior to the origin of land plants (vascular plants and bryophytes). However, the ancestor of all land plants was haploid-diploid.

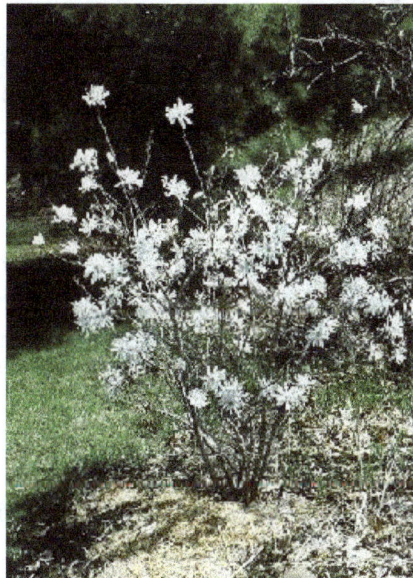
A vascular plant, Magnolia.

Haploid Life Cycle

Haploid life cycle is the most common in algae (without dikaryotic phase). Colonial alga such as red alga, brown alga, and green alga and filamentous alga such as red, green, and brown alga all exhibit haploid life cycle. A general life history includes both a 1n stage and a 2n stage separated by meiosis and syngamy. Before meiosis, the individual is diploidzygote. After the zygote goes through meiosis, it develops into a haploid (1n) spore or some other 1n structure. Meiosis produces four cells from each zygote and these four cells can be spores or other structure, depending on the organism. The four cells would then go through mitosis and become the organism. The organism has two options but not in all cases. It can either start asexual reproduction or it can produce gametes through mitosis. Isogamy and anisogamy, and sometimes oogamy may occur. The gametes then fuse in a process called syngamy, or fertilization. The fused gametes either all come from one single individual or from more than one individuals. After syngamy, the fused gametes become the zygote and become diploid again, the process repeats again. One important aspect about haploid life cycle is that only the gametophyte phase is present, the sporophyte does not exist in a haploid life cycle.

Haploid life cycle occurs in green algae. Volvox, for example is a colonial green algae in which both male gametes and egg are produced in the 1n stage, which then fuse together to form a zygospore, an encysted zygote that is protected from the harsh conditions of the environment. Another green algae that exhibits 1n life cycle is the Oedogonium, which is filamentous, or a chain of cells formed in one plane. Oedogonium grows in two different ways, through zoospores, or through syngamy of sperm and egg.

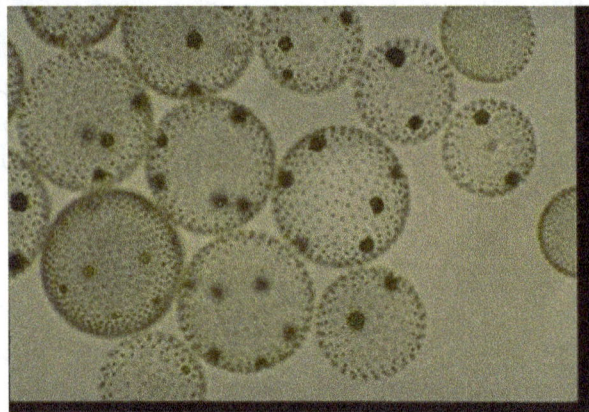

Volvox

In the first process, zoospores (mitospores) escape from the zoosporangium which is located in the parent algae and they develope into filaments. The parent also contains antheridia which produce sperm(1n) and an oogonium which produces the egg(1n). Syngamy occurs when the sperm and egg fuse and forms the zygote (2n). The 2n zygote then developes into the filamentous green algae.

Diploid Life Cycle

Contrary to the haploid life histories, sporophyte is present in a diploid life history and the gametophyte is absent. Meiosis produces 4 gametes (1n), ie. isogamy, anisogamy and oogamy, which through mitosis develops into mature gametes. These mature gametes then fuse (syngamy) and developes into zygote (2n). The gametes in syngamy can come from just one individual or different individuals. Through mitosis, the zygote develops into the organism. The organism then produces what is called a meiocyte (sporocyte), or other structure depending on the type of the organism and then go through meiosis to produce the 4 gametes and the cycle repeats itself. The organism can also go through asexual reproduction.

Fucus (rockweed), a brown alga, displays a diploid life history. The zygote (2n) becomes an embryo (very young sporophyte) and developes into the mature Fucus with receptacles at the tip of the algae. The receptacles are reproductive branches and contain many cavities with external pores which contain antheridia (male) and oogonia (female), then through meiosis, sperm (1n) and egg(1n) are produced. Syngamy occurs when the two fuse and become a zygote (2n). (Note that only the gametes are 1n.)

Fucus

Haploid - Diploid Life Cycle

The haploid-diploid life cycle is the most complex life cycle and thus has lots of variation. It is also the most common life cycle among plants since all land plants, the vascular plants and the bryophytes, are haploid-diploid. An alternation of generations defines the haploid-diploid, or 1n-2n, life cycle. This occurs when a multicellular 2n sporophyte (SPT) phase alternates with a multicellular 1n gametophyte (GPT) phase.

A bryophyte, Hornwort

This alternation of generations creates a morphologically complex life cycle, depending on the similarity or dissimilarity of the GPT and SPT, the relationship of these to each other, and the nature of the spores. There can be an alternation of similar generations, isomorphic alternation of generations, where the GPT and SPT are morphologically identical, except for the reproductive structures. On the other hand, there can also be an alternation of dissimilar generations, heteromorphic alternation of generations, when the GPT and SPT are morphologically different.

There are three possible relationships between the GPT and the SPT. The GPT and SPT may be totally independent of each other at maturity, independently free-living. This occurs in pteridophytes, or ferns, although the SPT stage of pteridophytes is much larger and is what we commonly recognize as the plant. The second type involves a GPT that is dominant over the SPT. The SPT is dependent of the GPT, as with all bryophytes. The most common relationship among land plants is when the SPT is dominant over GPT, as is the case with all vascular plants.

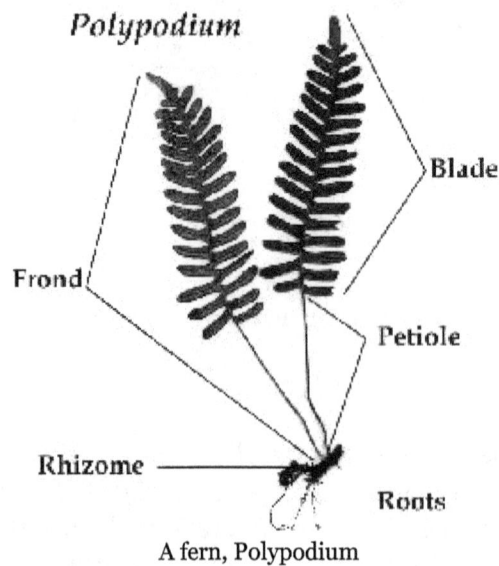

A fern, Polypodium

There are two main natures of spores. Some 1n-2n plants have only one morphological type of spore, and are called homosporous. Heterosporous plants have two morphologically different types of spores. They have male spores called microspores and female spores called megaspores. All GPT dominant plants are homosporous, whereas SPT dominant plants can be either homosporous or heterosporous.

References

- Cui L, Wall PK, Leebens-Mack JH, et al. (June 2006). "Widespread genome duplications throughout the history of flowering plants". GenomeRes. 16 (6):73849. doi:10.1101/gr.4825606. PMC 1479859. PMID 16702410.
- Ploidy, science: britannica.com, Retrieved 12 March 2018
- Comai L (November 2005). "The advantages and disadvantages of being polyploid". Nat. Rev. Genet. 6 (11): 836–46. doi:10.1038/nrg171110.1038/nrg1711. PMID 16304599.
- Haploid-plants, production-of-haploid-plants-with-diagram-10700: biologydiscussion.com, 12 April 2018
- Altergen, glossary: ucmp.berkeley.edu, 16 April 2018

Chapter 5

Genetically Modified Plants

Genetically modified crops are the plants that are used in agriculture for which the DNA has been modified using techniques of genetic engineering. A new trait is introduced in a plant or crop such as resistance to diseases, pests or environmental conditions, resistance to chemical treatments, reduction of spoilage, etc. This chapter discusses in detail the processes of gene gun method, protoplast isolation, microinjection, etc.

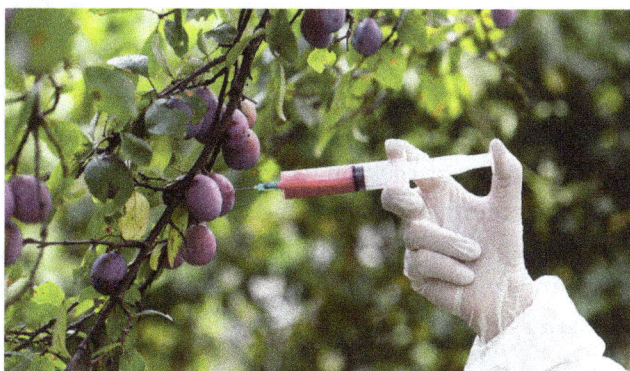

Genetically modified (GM) plants are also called transgenic plants. They are designed to acquire useful quality attributes such as insect resistance, herbicide tolerance, abiotic stress tolerance, disease resistance, high nutritional quality, high yield potential, delayed ripening, enhanced ornamental value, male sterility, and production of edible vaccines. Another major goal for raising the GM plants is their application as bioreactors for the production of nutraceuticals, therapeutic agents, antigens, monoclonal antibody fragments biopolymers, and so forth. Thus, GM plants can potentially affect many aspects of modern society, including agricultural production and medical treatment. Despite these potential applications, the use of GM plants for human welfare has been restricted owing to various concerns raised by the public and the critics. These concerns are divided into different categories, namely, health, nutritional, environmental, ecological, socioeconomic, and ethical concerns. These concerns include those arising due to properties of GM plants themselves, those resulting from the spread of the transgenes to other organisms, and also those resulting from their release into the environment. Such concerns have led to the withdrawal of commercialization of Bt cotton and Bt brinjal in India.

The campaign against GM plants was fueled by the instances of transgenic potatoes reported to be deleterious to rats, contamination of commercial corn products with unapproved StarLink and killing of monarch butterfly by Bt corn pollen. Furthermore, the nongovernmental organizations (NGOs) such as Gene Campaign, Center for Sustainable Agriculture, Research Foundation for Science, Technology and Ecology, Greenpeace, and Friends of the Earth have also raised concerns related to genetic manipulation of plants. The regulators, activists, media personnel and scientific journals have been undiscriminating and overly tolerant of the misrepresentations and distortions of anti-GM activists. There are not even scientific explanations for some of the concerns, but today the amount of misinformation is such that it has become difficult to separate truth from public perception about the GM plants. The biotechnology scientists, however, believe that GM plants should be given public acceptance because most of the concerns are not specific for GM plants and can exist for non-GM plants as well.

Concerns Related to Health and Nutritional Status

In case the products of GM plants are to be consumed by humans and animals, there is always a fear and risk in the society that these plants may create health problems or may lead to the development of newer microbial strains that may be pathogenic. Further, the plants themselves may be susceptible to such risks. Public and critics are also skeptical about the nutritional content and quality of the GM plants.

Susceptibility to Allergens

One of the major distressing problems with nontraditional proteins in GM foods is the risk of introducing allergens (usually glycoproteins) into the food supply of humans and animals. The public is concerned about the nature of these new food proteins as their allergenic or nonallergenic qualities are unknown. Allergenicity has been demonstrated in transgenic soybeans due to the transfer of a major food allergen from Brazil nuts. On the other hand, the scientists believe that the food allergens are found only in a few defined sources (peanut and other grain legumes, shellfish, tree nuts, etc.), and hence, only a dozen foods may produce allergic reactions. Moreover, allergenicity occurs when these food allergens are present in large proportions in the food and the individuals are sensitized to them over time to cause any adverse effects. Thus, it is highly unlikely for new allergens to be introduced into the food supply from GM plants.

Transfer of Antibiotic Resistance Gene to Microbes and Reduced Efficacy of Antibiotic Therapy

Public is also concerned about the potential risks associated with gene transfer from plants to microbes. It is speculated that the consumption of GM foods containing antibiotic resistance marker gene (e.g., Npt II gene encoding neomycin phosphotransferase for resistance to kanamycin and neomycin or gene encoding β-lactamase for resistance to ampicillin) by humans and animals may lead to transfer of these genes from GM food to microflora in the gut of humans and animals or to the pathogens in the environment transforming them into strains that are resistant to antibiotic therapy.

The transfer of antibiotic resistance gene to unrelated microorganisms such as Aspergillus niger has also been demonstrated. Biotechnology scientists, however, are of the opinion that the Npt II gene used to develop GM plants currently in the market is safe for use because there is no evidence of allergenicity or toxicity related to it. Moreover, humans are also susceptible to consuming several kanamycin resistant bacteria that occur naturally in the environment. Human gut is reported to contain 1012 kanamycin-resistant bacteria and by consuming a tomato harboring Npt II gene, the increase in frequency of kanamycin-resistant bacteria in the gut amounts only to 10–6%. Furthermore, acid conditions prevalent in stomach or rumen inactivate or degrade the encoded enzyme, neomycin phosphotransferase II.

Also neomycin phosphotransferase II requires ATP for its activity, which is present in extremely low concentrations in the gut. Regarding the use of gene for selection of bacterial recombinants, it is not transferred to plants. Moreover, the gene is considered safe because it does not encode for any product in plants. The Npt II and genes have been declared safe to use in GM plants. The public is, however, reluctant to accept this fact. Looking to the views of public, scientists have also developed nonresistance based selectable marker genes such as green fluorescent protein encoding gene (Gfp) and β-glucuronidase gene (Uid A). Besides, intron-containing Npt II gene has also been assessed as an efficient selectable marker in plant transformation. Due to insertion of intron in the Npt II gene, the theoretical risk of gene flow from GM plants to enteric bacteria is eliminated. Strategies for the removal of antibiotic resistance genes have also been devised.

One such strategy is the cloning of selectable marker gene and the transgene on two separate transfer DNA (T-DNA) molecules in a single plasmid or on two separate plasmids that are contained in one or more Agrobacterium tumefaciens strains used for plant transformation. The transgene and selectable marker gene are, thus, inserted at the loci, which should recombine at reasonably high frequencies so that the transgene can be segregated from the selectable marker gene in the next generation. Second strategy to eliminate the selectable marker gene is to flank it with direct repeats of recognition sites for a site-specific recombinase so that the marker gene can be easily excised from the plant genome by recombinase-mediated site-specific recombination. Examples included in this category are the Cre/lox recombination system of bacteriophage P1, Flp/frt recombination system of yeast 2 μm plasmid and R/Rs system of Zygosaccharomyces rouxii. A common feature of these systems is that the first round of transformation produces transgenic plants with the selection marker between two directly oriented recognition sites for the respective recombinase. After expression of recombinase, either by crossing in plants expressing the enzyme, by transient expression via second transformation, or by the use of an inducible promoter, the recombinase reaction is initiated resulting in marker-free transgenic plants. Marker gene may also be eliminated by placing it on a transposable element resulting in its loss after transposition.

The transgene by itself may be mobile and the activation of transposase allows the relocation of the desired transgene to a new chromosomal position. Genetic crosses and/or segregation may dissociate the two transgenes. Another novel strategy for the production of marker-free GM plants involves DNA deletion based on intrachromosomal homologous recombination between two homologous sequences, for example, by incorporating att sequence of λ bacteriophage.

Development of New-Line Microbial Strains

The third health risk is related to the ability of GM plants to create new toxic organisms. It is speculated that some nonpest microbial strains may acquire pathogenic trait by gene flow from GM plants. The risk can also be a new host being infected by a virus or recombining to form a more deadly virulent virus. Some plant pathologists also hypothesize that development of virus-resistant plants may allow viruses to infect new hosts through transencapsidation. Virus-resistant plants may also lead to the creation of new viruses through an exchange of genetic material or recombination between RNA virus genomes. Another matter of concern is that a small fraction of the DNA released from GM plants into soil may bind to the clay particles and hence protected from degradation. It is speculated that the soil bacteria may undergo transformation with the exogenous DNA of GM plant. This is, however, a rare possibility as the amount of DNA derived from GM plants as a proportion of the total DNA in the soil is likely to be very small, even if such plants are grown on a commercial scale. Moreover, the longevity of DNA in soil depends on various factors, including soil type and the presence of deoxyribonucleases in soil. target DNA from transgenic tobacco plants gets degraded within 40 days.

Skepticism About Nutritional Status

Critics of GM crops have raised various concerns about the potential of golden rice to combat vitamin A deficiency (VAD). The primary concern amongst these is the presence of insufficient vitamin A in golden rice. There are still doubts about the speed of degradation of vitamin A after harvesting the plant and the amount of vitamin A left after cooking. Vandana Shiva, an Indian anti-GMO activist, has criticized golden rice by saying "the golden rice is a hoax," "golden rice is a blind approach for blindness control," and "golden rice is just a recipe for creating hunger and malnutrition". She argues that the golden rice fails to pass the vitamin A need test and is incapable of removing VAD. It is calculated that one serving contains 30 g of rice on dry weight basis and golden rice can provide only 9.9 µg of vitamin A, that is, only 1.32% of the required daily allowance (RDA) of 750 µg. Even with the daily consumption of 100 g golden rice, only 4.4% of the required daily allowance will be met. Thus, an adult has to consume 2 kg 272 g of golden rice per day to complete his daily requirements of vitamin A. She is also of the view that, besides creating VAD, golden rice will also create deficiency in other micronutrients and nutrients.

This is because the raw milled rice has a low content of fat (0.5 g/100 g), which is necessary for vitamin A uptake, low content of protein (6.8 g/100 g), which is required as a carrier molecule, and low content of iron (0.7 g/100 g), which is required for the conversion of β-carotene to vitamin A. Friends of the Earth, Greenpeace, and Vandana Shiva further emphasize that there is no need of golden rice to combat VAD as superior alternatives such as sweet potato, green leafy vegetables, coriander, amaranth, carrot, pumpkin, mango, jackfruit exist in nature. It is reported that certain underutilized plants also exhibit far more nutritional value (vitamin A and other nutrients) than golden rice, for example, a combination of rice and leaves of Moringa (drumstick) tree, a native to India. Similarly, in contrast to rice, amaranth grain contains forty times more calcium, four times iron, and twice as much protein.

The ragi millet, grown in India, has thirty five times more calcium than rice, twice as much iron, and five times more minerals. It is opined that golden rice is not capable of increasing the production of β-carotene. Even if the target of 33.3 µg of vitamin A in 100 g of rice is achieved, it will be only 2.8% of β-carotene that can be obtained from amaranth leaves, 2.4% as that obtained from coriander leaves, curry leaves, and drumstick leaves. Thus, a far more efficient route to removing VAD is biodiversity conservation and propagation of naturally occurring vitamin A rich plants (wild-type or underutilized) in agriculture and diets. Even the World Bank has admitted that rediscovering and use of local plants and conservation of vitamin A rich green leafy vegetables and fruits have dramatically reduced VAD threatened children over the past 20 years in very cheap and efficient ways.

It is also speculated that the cultivation of golden rice will lead to major water scarcity since it is a water intensive crop and displaces water prudent sources of vitamin A. The scientists, on the other hand, believe that the traditional breeding methods have been unsuccessful in producing crops containing a high vitamin A concentration and most national authorities rely on expensive and complicated supplementation programs to address the problem. They also believe that a varied diet is beyond the means of many of the poor and they have to rely on one or few foods to provide complete nutrition, for example, rice.

Thus, golden rice may be a useful tool to help treat the problem of VAD in young children living in the tropics. They also emphasize that the critics are ignoring the fact that VAD disorders result from a deficiency of vitamin A and not its complete absence in the diet and the VAD individuals lack only 10%–50% of their daily requirements. Hence, any additional contribution toward daily requirements would be useful. In 2005, a team of researchers at Syngenta have produced a variety of golden rice, called "Golden rice 2," which produces twenty-three times more carotenoids than golden rice (up to 37μg/g) and preferentially accumulates β-carotene (up to 31μg/g of the 37μg/g of carotenoids). The Rockefeller Foundation emphasized that the new strains of golden rice contain substantially higher levels of β-carotene than the early versions on which the opponents based their calculations. In order to meet the RDA, 144 g of the most high-yielding golden rice strains would have to be eaten.

Environmental and Ecological Concerns

Large-scale cultivation of GM plants expressing viral and bacterial genes and their release into the environment is considered to be a threat and called as "genetic pollution" by the critics. The risk of a transgene spreading in the environment is related to the likelihood for out-crossing, horizontal gene transfer, and the phenotype imparted by the gene. Debates about the commercial introduction of GM plants in some parts of the world have led to questions about their potential impact on the environment unless necessary safeguards are taken into account. It should, however, be acknowledged that agriculture inevitably has an impact on the environment and these concerns are not specific for GM plants.

Transgene Escape to Wild-Type Plants

There is a potential risk that the GM plants may hybridize (or cross-breed) with sexually compatible wild-type species. This genetic exchange is possible due to wind pollination, biotic pollination or seed dispersal. This may have an impact on the environment through the production of hybrids and their progeny. In an example, virus-resistant squash commercialized in 1994 was demonstrated to transfer its virus resistance gene to wild squash (Cucurbita pepo), an agricultural weed native to the southern United States, thereby decreasing its value to squash breeders. On the other hand, it is significant to note that for an effective pollen transfer to occur, the GM plants must be close enough to the wild species, should flower at same time, and must be genetically compatible. Further, the risk of any gene transfer to related weedy species through pollen has been eliminated by devising chloroplast transformation procedures. This is because, in many crop species, chloroplasts display only maternal inheritance.

Selective Advantage to GM Plants in Natural Environments and Generation of Super weeds

The concern of gene flow from GM plants to weedy relatives via pollination is quite intense. It is considered that the transfer of encoded characteristics to weed species could potentially give them a selective advantage, consequently leading to the generation of "superweeds." Moreover, the newly introduced traits may make a plant, especially herbicide tolerant plant, more persistent or invasive (weedy) in agricultural habitats. It is, however, pertinent to note that the risk of gene transfer to weeds is similar with both conventional and GM plants and is not contingent on how these genes have been introduced into plants. Such a risk of gene flow has always existed since the advent of modern plant breeding, even when there were no GM plants, and this can occur where possible. Several studies have demonstrated that tolerance to particular herbicide is often more likely to develop by evolution from within the weed gene pool rather than by gene flow from herbicide-tolerant plants. Nevertheless, the current scientific evidence indicates that the weediness arises from many different characters and that the addition of one gene is unlikely to cause a crop to become a weed. The transfer of novel genes from transgenics (or even conventionally bred plants) to weeds depends on the nature of the novel gene and the biology and ecology of the recipient weed species. The probability of successful out-crossing thus depends on sexual compatibility, physical proximity, distance of pollen movement both out of and into the GM plants, and ecology of recipient species. Thus, only a few plants such as oilseed rape, barley, wheat, beans, and sugar beet can hybridize with weeds. For example, oilseed rape has been reported to hybridize with hoary mustard, wild radish, and other wild Brassica species. Furthermore, the transfer of herbicide tolerance gene is unlikely to confer any competitive advantage to hybrids outside agricultural areas. It is also comforting to recognize that there is no proven evidence of enhanced persistence or invasiveness of GM plants and no major super weeds have developed so far.

Effect on Nutritional Composition of Plants

It is also speculated that the nutritional composition of GM products may be affected in GM plants. Another concern is that the transgenes from animals (obtained from fishes, mouse, human, and microbes) introduced into GM plant for molecular farming may pose a risk of changing the fundamental nature of vegetables. In a study, it was reported that as compared to non-GM soybean, GM plants exhibited lower levels of isoflavones . This finding also raised a doubt on the regulatory system for the release of the GM plants. However, later it was found that the concentration of isoflavone in GM soybean was within the normal range.

Mixing Genes from Unrelated Species (Interbreeding)

The public is worried about the risk that the GM plants can spread through nature and interbreed with natural organisms, thereby contaminating "non-GM" environments. This would in turn affect the future generations in an unforeseeable and uncontrollable way . Such worries, however, ignore the history of plant breeding and the existing overwhelming sequence similarity of genes across kingdoms.

Development of Tolerance to Target Herbicide

It is viewed that the repeated use of the same herbicide in the same area to remove weeds amongst genetically modified herbicide-resistant crops (HRCs) (tolerant to single herbicide) will exacerbate the problem of herbicide-tolerant weeds . Another matter of concern relates to the plants carrying different herbicide tolerance genes to become multiply tolerant to several herbicides by pollination between adjacent plants . In several closely studied examples in Canada, farmers have detected oilseed rape plants tolerant to three different herbicides (note that two were acquired from GM plants and the third possibly from conventional breeding) . The development of multiple tolerances in "volunteer" crop plants (from seeds remaining viable in agricultural soil) may also exert an impact on the environment by necessitating the use of less environment-friendly and possibly outdated herbicides by the farmers. On the other hand, the proponents believe that herbicide resistance develops due to excessive application of herbicide and is not exclusively associated with gene transfer from genetically modified HRCs. Thus, the pressure on weeds to evolve resistant biotypes has been reported to be pronounced with the excessive application of herbicides such as glyphosate, sulphonylureas, and imidazolinones.

Sustainable Resistance in Insect Pests

It is possible that the widespread use of disease-resistant GM plants may lead to the evolution of several insect pests that are resistant to pesticides. For example, Bt crops may develop resistance to Bt biopesticide, a permitted biopesticide successfully used by organic farmers in the integrated pest management (IPM) programs. There is to date no reported evidence of insect resistance to Bt crops under field conditions although Bt resistant insects (e.g., cotton budworm and bollworm) have been observed in areas where Bt biopesticides are sprayed on crops . It has been a matter of concern that the development of such resistance may lead to the loss of the potential of the Bt biopesticide, which may in turn make it necessary for organic farmers to resort to less environmentally acceptable chemical pesticides. Therefore, proper resistance management strategies along with this comparatively newer technology are imperative. The most widely used is the 'high-dose refuge' strategy designed to prevent or delay the emergence of Bt toxin-resistant insects. Scientists are of the opinion that this strategy should be followed without fail, as the rate of noncompliance can increase the risk of plant resistance breakdown.

Harm to Nontarget Organisms

Nontarget effect, that is, undesirable effect of a novel gene (usually conferring pest or disease resistance) on "friendly" organisms in the environment, is another concern related to GM plants . As many nontarget microbes harbor on plant surfaces or some insects harbor on flowers, it becomes quite challenging to target the insect resistance gene product to appropriate plant tissues and hence kill pests without exerting any adverse effect on friendly organisms such as pollinators and biological control agents. This is particularly difficult where the benign or beneficial organism is related and physiologically similar to the pest to be targeted. One of the most significant studies of nontarget impacts of GM plants has been the killing of monarch butterfly in the United States by Bt insecticidal proteins. It should, however, be noted that the pesticidal sprays used on Bt or non-Bt corn may be more harmful to the monarch butterfly as compared to Bt corn pollen. Thus, in evaluating the use of Bt crops and the possible environmental damage caused, it is important to take into account the environmental damage caused by the use of pesticides in agriculture generally. It is argued that millions of birds and billions of insects, both harmful and beneficial, are killed each year due to excessive use of pesticides. It is, however, suggested that the scale and pattern of use may mitigate the effects of Bt on nontarget populations. Furthermore, when toxins are produced within plant tissues, nontarget organisms are exposed to a much lesser extent than with spray applications because only those organisms which feed on the plant tissues come into contact with the toxin.

Harmful effect of Bt toxin residues in the soil after harvest of the GM crop on soil invertebrates has been another matter of concern. An investigation of the effect of Cry1Ab released from the roots and crop residues on soil organisms revealed the presence of toxin in the guts and casts of tested earthworms. There was, however, no significant difference in their mortality or weight. Moreover, no difference in the total number of other soil organisms (including nematodes, protozoa, bacteria, and fungi) between the soil rhizosphere of Bt and non-Bt crops was detected .

Increased Use of Chemicals in Agriculture

On one hand, the transgenes conferring herbicide resistance have been criticized because these would maintain, if not promote, the use of herbicides and their attendant problems [125, 126]. Similarly, there is a concern that the insect-resistant and disease-resistant GM plants will increase the application of insecticides and pesticides, respectively. On the contrary, reports demonstrate that there is no significant change in the overall amount of herbicide use in the United States since the introduction of GM soybeans . An analysis by soybean growers at the United States has shown that $7.2 millions of other herbicides were replaced by $5.4 millions of glyphosate . This substitution, thus, resulted in the replacement of highly toxic and more persistent herbicides with that of glyphosate. Furthermore, it has been reported that herbicide-tolerant oilseed rape eliminates the use of >6,000 tons of herbicide in the growing season .

Loss of Biodiversity

The public has long been worried about the loss of plant biodiversity due to global industrialization, urbanization, and the popularity of conventionally-bred high-yielding varieties. It is speculated that the biodiversity will be further threatened due to the encouraging use of GM plants. This is because development of GM plants may favor monocultures, that is, plants of a single kind, which are best suitable for one or other conditions or produce one product. Further, the transformation of more natural ecosystems into agricultural lands for planting GM plants is adding to this ecological instability.

Another point of concern is the loss of weed diversity that may occur due to gene flow from HRCs to weeds . It is argued that because the currently available HRCs confer tolerance to broad-spectrum herbicides such as glufosinate and glyphosate, their extensive use may shift the diversity of weeds in agricultural habitats. However, weeds exhibit considerable plasticity and adapt to a wide range of cultivation practices. Experience with conventional agriculture has shown that weed species composition varies within the same crop among different fields and at different times of year. Thus, weed population shifts are natural ecological phenomena in crop management and should not be viewed as exclusive to GM plants.

Unpredictable Gene Expression

It is speculated that the random gene insertion, transgene instability, and genomic disruption due to gene transfer may result in unpredictable gene expression. Such a risk is, however, unlikely to be unique to GM plants or of any significance considering our current knowledge of genomic flux in plants.

Alteration in Evolutionary Pattern

Plants adapt to the fluctuations in the environment through changing their genes and developing better races called "evolved races." These mutations, however, occur at a very low frequency (i.e., one in about 109/gene/generation). It is hypothesized that the cultivation of GM plants by the farmers at an increasing rate throughout the world may change the evolutionary pattern drastically . Another concern is the evolution of non-GM plants through hybridization with GM plants.

Loss of Ecosystem in Marginal Lands

As new plants are introduced mainly to marginal lands, loss of natural ecosystems in these areas has also been a matter of concern.

Contamination of Soil and Water

It is also sometimes argued that the widespread introduction of HRCs will increase the use of herbicides, which will in turn contribute to the contamination of soil and ground water. However, this is not the case. The cultivation of HRCs in the United States has been reported to facilitate zero-till agronomic system, which contributes to a reduction in soil erosion. The release of Bt toxin into the soil after harvest of Bt crops is also viewed as a risk factor associated with the cultivation of Bt crops [123, 124, 130]. It has been found that Bttoxins remain active in soil; however, it is not necessarily an environmental hazard because Bt toxins must be ingested and affect only selected groups of insects. Moreover, the potential leaching rate of Bt toxin is reduced due to its binding and adsorption on clay particles.

Genetically Modified Crops

The term genetically modified (GM), as it is commonly used, refers to the transfer of genes between organisms using a series of laboratory techniques for cloning genes, splicing DNA segments together, and inserting genes into cells. Collectively, these techniques are known as recombinant DNA technology. Other terms used for GM plants or foods derived from them are genetically modified organism (GMO), genetically engineered (GE), bioengineered, and transgenic. 'Genetically modified' is an imprecise term and a potentially confusing one, in that virtually everything we eat has been modified genetically through domestication from wild species and many generations of selection by humans for desirable traits. The term is used here because it is the one most widely used to indicate the use of recombinant DNA technology.

Traits that have been Modified in GM Crops

Insect-resistant crops contain genes from the soil bacterium Bacillus thuringiensis (Bt). The protein produced in the plant by the Bt gene is toxic to a targeted group of insects—for example European corn borer or corn rootworm—but not to mammals. The most common herbicide tolerant (HT) crops are known as Roundup Ready, meaning they are tolerant to glyphosate (the active ingredient in Roundup herbicide). Glyphosate inactivates a key enzyme involved in amino acid synthesis that is present in all green plants; therefore, it is an effective broad spectrum herbicide against nearly all weeds. Roundup Ready crops have been engineered to produce a resistant form of the enzyme, so they remain healthy even after being sprayed with glyphosate. Some cultivars of corn and cotton are referred to as 'stacked', meaning they have transgenes for both insect resistance and HT.

Potential GM Crops of the Future

Some potential applications of GM crop technology are:

- Nutritional enhancement: Higher vitamin content; more healthful fatty acid profiles;

- Stress tolerance: Tolerance to high and low temperatures, salinity, and drought;

- Disease resistance: For example, orange trees resistant to citrus greening disease or American chestnut trees resistant to fungal blight;

- Biofuels: Plants with altered cell wall composition for more efficient conversion to ethanol;

- Phytoremediation: Plants that extract and concentrate contaminants like heavy metals from polluted sites.

Difference between GM Technology and Plant Breeding Techniques

The era of scientific crop improvement dates back to around 1900, when the impact of Gregor Mendel's studies on trait inheritance in peas became widely recognized. Since then, a broad range of techniques has been developed to improve crop yields, quality, and resistance to disease, insects, and environmental stress. Most plant breeding programs rely on manual cross-pollination between genetically distinct plants to create new combinations of genes. The progeny plants are intensively evaluated over several generations and the best ones are selected for potential release as new varieties. An example is a tomato variety that is selected for disease resistance and tolerance to cool temperatures. Other techniques included within the conventional plant breeding toolbox are development of hybrid varieties by crossing two parental strains to produce offspring with increased vigor; and induced mutations to create useful variation. GM technology is much more precise in that it transfers only the desired gene or genes to the recipient plant. Another branch of agricultural biotechnology—distinct from GM technology—involves selecting plants for DNA patterns known to be associated with favorable traits such as higher yield or disease resistance.

The Shared DNA Code

Most organisms store their genetic information in the form of DNA molecules in chromosomes. The sequence of chemical bases in a DNA strand encodes a specific order of amino acids, which are the building blocks of proteins. Proteins carry out many functions in cells and tissues, which together are responsible for an organism's characteristics. Because most life forms share this same language of heredity—and due to scientific advances in molecular biology—it is now possible to transfer a gene from one species to another, for example from a bacterium to a plant, and have it function in its new host.

Types of Genetically Modified Plants

Transgenic Plants

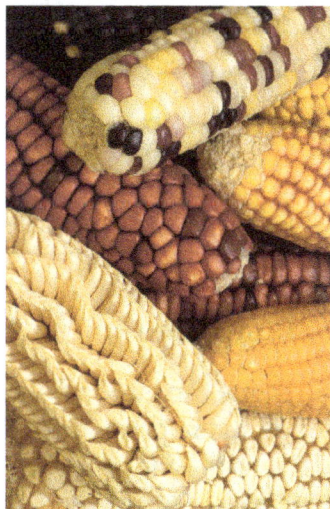

Transgenic plants or genetically modified plants are plants whose DNA is modified using genetic engineering techniques. In most cases the aim is to introduce a new trait to the plant which does not occur naturally in this species. Examples include resistance to certain pests, diseases or environmental conditions, or the production of a certain nutrient or pharmaceutical agent.

A transgenic crop plant contains a gene or genes which have been artificially inserted instead of the plant acquiring them through pollination. The inserted gene sequence, known as the transgene, may come from another unrelated plant or from a completely different species. Plants containing transgenes are often called genetically modified or GM crops.

Reasons for Making Transgenic Crop Plants

- The process of transgenic plant development primarily aims at assembling a combination of genes in a crop plant which will make it as useful and productive as possible. Depending on where and for what purpose the plant is grown, desirable genes may provide features such as higher yield or improved quality, pest or disease resistance, or tolerance to heat, cold and drought.

- Combining the best genes in one plant is a long and difficult process, especially as traditional plant breeding has been limited to artificially crossing plants within the same species or with closely related species to bring different genes together.

   For example, a gene for protein in soybean could not be transferred to a completely different crop such as corn using traditional techniques. Transgenic technology enables plant breeders to bring together in one plant useful genes from a wide range of living sources, not just from within the crop species or from closely related plants.

- This technology provides the means for identifying and isolating genes controlling specific characteristics in one kind of organism, and for moving copies of those genes into another quite different organism, which will then also have those characteristics.

- This powerful tool enables plant breeders to do what they have always done-to generate more useful and productive crop varieties containing new combinations of genes-but it expands the possibilities beyond the limitations imposed by traditional cross-pollination and selection techniques.

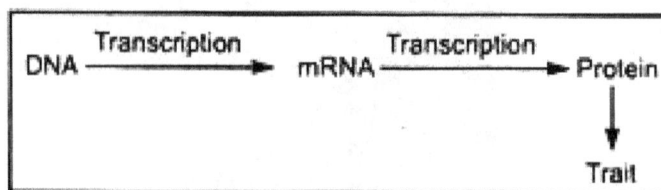

Fundamentals of Transgenic Plant Development

The underlying reason that transgenic plants can be constructed is the universal presence of DNA (deoxyribonucleic acid) in the cells of all living organisms. This molecule stores the organism's genetic information and orchestrates the metabolic processes of life. Genes are discrete segments of DNA that encode the information necessary for assembly of a specific protein.

A specific protein (or an enzyme) encodes for a particular trait. In the production of a transgenic plant our primary aim is to transfer a foreign gene, encoding for some novel traits, into the genome of the plant stably.

After transferring we also need the transgene to integrate and express in the plant's cells. This process as a whole generates a new variety of plant which is new in its own kind and interests us in its massive large scale cultivation.

Steps Involved in the Production of Transgenic Plants

The fundamental steps involved in the transgenic plant production are as follows:

Step: Identifying, Isolation and Cloning of Genes for Agriculturally Important Traits

The very first step in the generation of transgenic plant is to identify and isolate the novel transgene that we want to transfer into the genome of the target plant. Usually, identifying a single gene involved with a trait is not sufficient. We also have to understand how the gene is regulated, what other effects it might have on the plant, and how it interacts with other genes active in the same biochemical pathway.

Step: Designing Gene Construct for Insertion

After entering the plant cell the transgene must inter-grate into the genome of the plant stably and express itself successfully so as to produce higher amount of transgenic protein which will be indirectly reflected in the trait controlled by it.

To achieve this we have to design a "gene construct" or gene-set, having all the DNA segments necessary to achieve the integration and expression of the transgene. Once a gene has been isolated and cloned (amplified in a bacterial vector), it must undergo several modifications before it can be effectively inserted into a plant.

A gene-set which will be transferred to the target plant has following segments:

A Promoter Sequence

This must be added for the gene to be correctly expressed (i.e., translated into a protein product). The promoter is the on/off switch that controls when and where in the plant the gene will be expressed. To date, most promoters in transgenic crop varieties have been "constitutive", i.e., causing gene expression throughout the life cycle of the plant in most tissues.

The most commonly used constitutive promoter is CaMV 35S, from the cauliflower mosaic virus, which generally results in a high degree of expression in plants. Other promoters are more specific and respond to cues in the plant's internal or external environment. An example of a light-inducible promoteris the promoter from the cab gene, encoding the major chlorophyll a/b binding protein.

The Transgene

Sometimes, the transgene is modified to achieve greater expression in a plant.

For example, the Bt gene for insect resistance is of bacterial origin and has a higher percentage of A-T nucleotide pairs compared to plants, which prefer G-C nucleotide pairs. In a clever modification, researchers substituted A-T nucleotides with G-C nucleotides in the Bt gene without significantly changing the amino acid sequence. The result was enhanced production of the gene product in plant cells.

Termination Sequence

This signals to the cellular machinery that the end of the gene sequence has been reached.

A Selectable Marker Gene

This is added in order to identify plant cells or tissues that have successfully integrated the transgene. This is necessary because achieving incorporation and expression of transgenes in plant cells is a rare event, occurring in just a few per cent of the targeted tissues or cells.

Selectable marker genes encode proteins that provide resistance to agents that are normally toxic to plants, such as antibiotics or herbicides. Only plant cells that have integrated the selectable marker gene will survive when grown on a medium containing the appropriate antibiotic or herbicide.

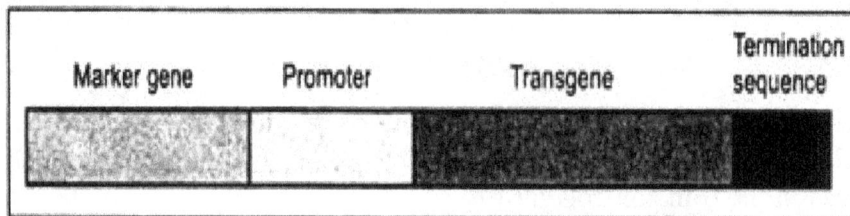

The essential features of an ideal reporter gene are:

1. Lack of endogenous activity in plant cells of the concerned enzyme,

2. An efficient and easy detection, and

3. A relatively rapid degradation of the enzyme.

The commonly used selectable marker genes include those conferring resistance to the antibiotics kanamycin (nptII, encoding neomycin phosphotransferase) and hygromycin (hptIV, encoding hygromycin phosphotransferase, isolated from E. coli); and broad range herbicides glyphosate (modified versions of the enzyme EPSPS, 5-enolpyruvate shikimate-3-phosphate synthase, isolated from E. coli or Salmonella typhimurium), phosphinothricin (bar, isolated from Streptomyces hygroscopicus, codes for phophinothricin acetyltransferase), etc.

Step: Transforming Target Plants with the Gene Construct

There are two ways of genetically transforming the target plant:

1. Vector mediated gene transfer, and

2. Vector less or direct gene transfer.

Step: Selection of The Transgenic Plant Tissue/Cells

Following the gene insertion process, plant tissues are transferred to a selective medium containing an antibiotic or herbicide, depending on which selectable marker was used. Only plants expressing the selectable marker gene will survive and it is assumed that these plants will also possess the transgene of interest.

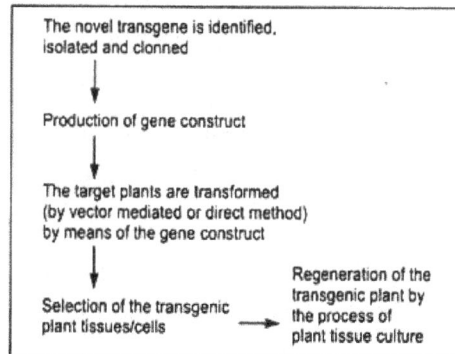

```
The novel transgene is identified,
isolated and clonned
        |
        v
Production of gene construct
        |
        v
The target plants are transformed
(by vector mediated or direct method)
by means of the gene construct
        |                      Regeneration of the
        v                      transgenic plant by
Selection of the transgenic  ----->  the process of
plant tissues/cells            plant tissue culture
```

Step: Regeneration of the Transgenic Plants

To obtain whole plants from transgenic tissues, they are grown under controlled environmental conditions in a series of media containing nutrients and hormones by the process of plant tissue culture.

Promoters used in construction of gene construct

Promoter	Source	Relative Activity
35S	CaMV 35S RNA gene	Constitutive, high activity; most commonly used in dicots
35S + Adh1-I 1	35S promoter + first intron of maize Adh1 gene	Enhanced promoter activity; constitutive
35S + sh1-I 1	35S promoter + first intron of maize shrunken 1 gene	Better than 35S + Adh1-I 1 in mono-cots; constitutive
Adh1	Promoter of alcohol dehydrogenase gene of maize	Moderate activity in cereals; anaerobic expression
Emu	Modified from Adh1 promoter and its first intron	Moderate activity in cereals; anaerobic expression
Act1 + Act-I 1	Rice actin gene + its first intron	Moderate activity; constitutive
Ubi1 + Ubi1-I 1 (or I 6)	Maize ubiquitin 1 gene promoter + its first (or sixth) intron	High activity in cereals; constitutive
Vicilin promoter	Pea vicilin storage protein gene	Seed-specific promoter

Integration of the Transgene in the Genome of the Target Plant

In general, transgenes integrate at random sites in any of the chromosomes of the genome of host cells. Usually, in a given cell, integration occurs at a single location. As a result, different cells may be expected to show integration of the transgene at different chromosomal locations.

The number of copies integrated per genome ranges from one to several hundred. In general, multiple copies are integrated when large amounts of DNA are used for transfection, while single copies are integrated with smaller amounts.

When multiple copies are integrated, they are mostly integrated at one site joined to each other head-to-tail, i.e., as a concatemer. However, in a small proportion of cases, the multiple copies are located at several sites in the same genome.

The mechanism of random integration is not known. The entire gene construct, including the vector DNA, becomes integrated. When two different gene constructs are mixed and used for transfection, they tend to be integrated together at the same site; this is known as co-transfection. The sequences flanking a gene on either side influence the expression of this gene.

Therefore, the same transgene integrated at different locations in the genome may show different levels of expression; this is known as position effect. Transgene integration frequently leads to various forms of rearrangements, e.g., duplication, deletion, etc., near the site of integration.

If these changes are large enough, the host gene located at the site of integration may become non-functional. A host gene would also become non-functional if the transgene becomes integrated within the coding region of this gene. When integration of a transgene leads to the loss of function of a host gene, it is called insertional mutagenesis; it often produces aberrant phenotypes.

Analysis of Transgene Integration

The integration of transgene into the genome is confirmed by Southern hybridization of genomic DNA extracted from the considered transgenic individuals. The DNA is digested with a suitable restriction enzyme prior to electrophoresis.

By choosing appropriate restriction enzymes for DNA digestion, not only the integration of transgene can be established beyond doubt, but information on the number of copies per cell, the orientations of tandemly arranged copies and the presence of single or multiple integration sites is also obtained from Southern hybridization. All the individuals that give positive result with Southern hybridization are regarded as confirmed transgenic.

Detection of mRNA Expression

The mRNAs produced by transgenes is most readily detected if they are with unique sequences, which have no counterparts among those produced by the host genome. A high purity RNA preparation is obtained from the appropriate tissue of transgenic individuals, and is subjected to RNA dot blot hybridization with a radioactive probe specific for the transgene.

Alternatively, the RNA preparation may be used for northern hybridization, which provides additional information on transcript size as well.

Inheritance of Transgenes

The transgenes which are stably integrated are inherited in a Mendelian fashion. They are usually dominant. Instability may occur due to point mutation, like methylation, or rearrangements of the T-DNA region. In addition, homologous recombination between copies of the transgene inserted in the same nucleus can also lead to instability of the gene.

Future Development of Transgenic Technology

New techniques for producing transgenic plants will improve the efficiency of the process and will help resolve some of the environmental and health concerns.

Among the expected changes are the following:s

- More efficient transformation, that is, a higher percentage of plant cells will successfully incorporate the transgene.

- Better marker genes to replace the use of antibiotic resistance genes.

- Better control of gene expression through more specific promoters, so that the inserted gene will be active only when and where needed.

- Transfer of multi-gene DNA fragments to modify more complex traits.

Cisgenic Plants

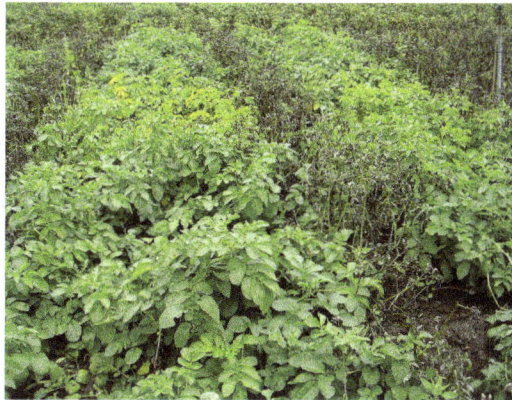

Cisgenic plants are made using genes found within the same species or a closely related one, where conventional plant breeding can occur. Some breeders and scientists argue that cisgenic modification is useful for plants that are difficult to crossbreed by conventional means (such as potatoes), and that plants in the cisgenic category should not require the same regulatory scrutiny as transgenics.

Cisgenesis is the production of genetically modified crops using a donor DNA fragment from the species itself or from a cross compatible species. The newly introduced gene is unchanged and includes its own introns and regulatory sequences and is free of any vector DNA, except T- DNA border sequences that flank the cisgene. The resultant phenotype of the cisgenic plant can be achieved through conventional breeding also; but it will take a much longer time. One of the most important points of cisgenesis is that it introduces only the desired gene, thus avoiding linkage drag that can be resulted from conventional cross breeding and also it eliminates hectic and time consuming backcrossing to recover the recurrent parent genotype. The concept of cisgenesis was introduced by Dutch researchers; Schouten, Krens and Jacobsen in the year 2006.

The worthiness of the GM techniques for developing highly reliable and good quality food products to the world has been set off by public worries about the safety of the derived food and there subsequent finished products. Most particularly, the controversy has spotlighted on the probable unpredictable hazards arising from the agglomeration of certain new substances in crop plants that confers toxicity, allergy and genetic threats to humans.

Benefits of Cisgenesis

- Cisgenesis is as safe as conventional breeding.

- To overcome the problem of linkage drag.

- Original genetic makeup of a crop is preserved any one or few genes are added.

- Escape of foreign genes via pollen grains to natural vegetation is not a problem in cisgenesis.

- In cisgenesis, only the desired genes are introduced without the undesirable genes unlike conventional breeding.

Drawbacks of Cisgenesis

- The gene outside the sexually compatible gene pool cannot be introduced.

- There is also a chance that the introduction of cisgene may influence the expression of genes that are already present in the recipient genome.

- Position effect may lead to alteration of the gene expression and phenotypic differences.

- The production of marker-free plants often requires the implementation or development of new techniques and such techniques may not be readily available.

Advantages of Cisgenesis over Conventional Breeding

1. Removes linkage drag: Introgression of new traits into the cultivated varieties by conventional methods comprises of wide crosses and repeated backcrossing. However, these traits are constantly linked with a large share of unwanted chromosomal region, the so called linkage drag. Some of these genes affect the normal features of the crop as they may engage in the production of diverse kinds of toxins or allergens. Hence, direct transfer of desired genes through cisgenesis into an existing variety without altering any of the properties enviable for the consumers can be accomplished.

2. Maintains the original genetic make-up of the plant varieties: In a hybridization method, the genetic makeup of the progeny plant varies from its parents because it has been a mixture of both the parental genomes. In spite of this, there is a necessity to conserve some part of the genome which reveals certain constructive traits. Through conventional plant breeding such an approach is not possible entirely due to self incompatibility among the vegetatively propagated plants. Cisgenic breeding tools are used in order to get up to four different resistance genes into one variety without changing the other traits of the modified variety.

3. Reduction in pesticide application: The key purpose of cisgenesis is to transfer disease resistance genes to susceptible varieties. The vital goal here is to lessen substantial pesticide application. As a result, there is decline in the input costs of the farmers and decreased pesticide leftovers on the plants and also in their products, which is mostly favoured by the consumers. This reduces the environmental pollution by pesticides and in turn helpes in sustainable agricultural development.

4. Time Saving: In conventional hybridization programmes, there is linkage drag, where there is inheritance of unwanted genes to the progeny. Several backcrossed generations are required to get rid of such kind of undesired genes. Cisgenesis overcomes the problem of linkage drag and only the gene of interest is introduced into the genome of the recipient plant within a short period of time. Thus, this saves a lot of time.

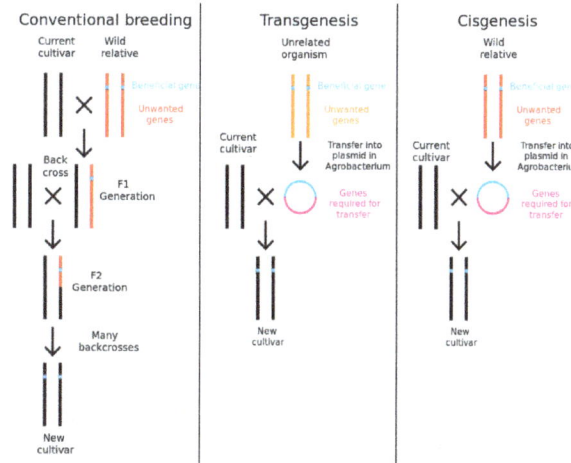

Subgenic

Genetically modified plants can also be developed using gene knockdown or gene knockout to alter the genetic makeup of a plant without incorporating genes from other plants. In 2014, Chinese researcher Gao Caixia filed patents on the creation of a strain of wheat that is resistant to powdery mildew. The strain lacks genes that encode proteins that repress defenses against the mildew. The researchers deleted all three copies of the genes from wheat's hexaploid genome. Gao used the TALENs and CRISPR gene editing tools without adding or changing any other genes. No field trials were immediately planned. The CRISPR technique has also been used to modify white button mushrooms (Agaricus bisporus).

Multiple Trait Integration

From a breeding standpoint, multiple trait integration (MTI) is a four-step process of converting an elite variety/hybrid for value-added traits (e.g. transgenic events) using backcross breeding, ultimately regaining the performance attributes of the target hybrid along with reliable expression of the value-added traits.

Protoplast Isolation

Each plant cell has a definite cellulose wall which is attached with each other by middle lamella made up of pectin. Protoplast is the content which is surrounded only by plasma-lemma. So to get the isolated protoplast it is essential to remove the pectin material to obtain the single cells and then to remove cell wall by enzymatic method.

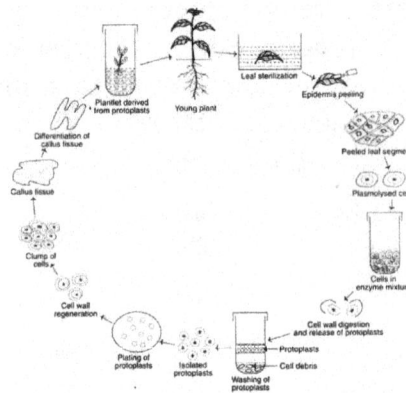

Sources of Explant for Protoplast Isolation

Protoplasts can be isolated directly from the different parts of whole plant which bears the soft parenchymatous tissue (e.g., young fully expanded soft leaves) or indirectly from the in vitro grown plant tissue (e.g., callus tissue). Protoplast yield and viability depends greatly on the type of tissue material from which it is isolated and the method used. Cell suspension cultures may also provide a very good source of protoplast.

Before isolation of protoplast, the source material if it is from in vivo grown plant then it should be properly surface sterilized using the proper method of sterilization. Then any of the methods either mechanical or enzymatic can be used to isolate the protoplast.

Methods of Protoplast Isolation

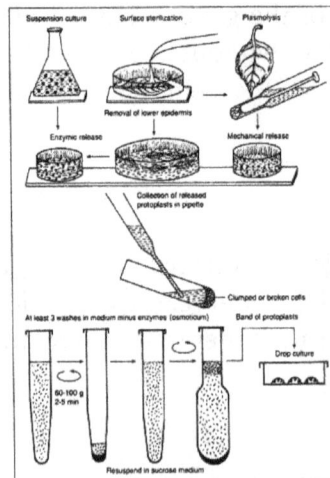

The essential step of protoplast isolation is the proper use of osmoticum. The plant protoplast is an osmotic system where the plasma membrane acts as the semipermeable membrane which equilibrates the outward and inward flow of water. If the inward pressure is more than due to heavy inflow of water the protoplast will burst.

So during protoplast isolation the enzymes applied to isolate the protoplast and the tissue material should be placed in proper osmolyticum or plasmolyticum. Generally mannitol, a sugar alcohol at high concentration, an easily transportable component provides a good stable osmotic environment which prevents bursting of protoplast.

Protoplasts may be isolated by any one of the two following ways:

- Mechanical method (Non-enzymatic method);
- Enzymatic method.

Mechanical Method

Any soft parenchymatous tissue is kept in a plasmolyticum. The plasmolysed tissue is then finely chopped into pieces and the intact cells (plasmolysed) are released into the medium from the cut surface. The suspension is then allowed for deplasmolysis and the released protoplasts attain their original size. The yield of protoplast in this method is very low, for large scale of protoplast yield the enzymatic method is followed.

Enzymatic Method

Young fully expanded soft leaves, or in vitro grown callus tissue or cell suspension culture grown cells can be used as the source material. The tissues or cells are incubated in plasmolyticum for 1 hr before enzymatic treatment. The intact tissue materials cut into smaller pieces to increase the surface area of enzymatic activity. The enzymes can be used either sequentially in two step method or in a single step by mixed enzymatic method.

The enzymes used are of three main categories:

- Cellulases

  Cellulysin

  Driselase

- Hemicellulase

  Helicase

  Rhozyme

- Pectinase

  Macerozyme-R-10

  Pectinol

The concentration of enzymes used and the time period of incubation varies greatly depending on the tissue type. In two step method, the pectinase and hemicellulases are applied first and then cellulase is applied for complete removal of cell wall.

After release of protoplast into the suspension, for removal of enzymes the protoplasts are collected in centrifuge tube as pellet and washed several times with the osmoticum.

Purification of Protoplasts

For purification, the protoplasts suspended in osmoticum are centrifuged using sucrose (20%) solution. The viable protoplasts float on the top surface of sucrose solution forming a band. These protoplasts are then collected, re-suspended in osmoticum and washed several times.

Finally the protoplasts suspended in a measured volume of protoplast culture medium after counting the number with the help of hemocytometer. The viability of protoplast is checked with the help of Fluorescin diacetate staining or phenosafranine or calcofluore white.

Isolation of Sub-Protoplasts

Sub-protoplasts do not contain the entire content of plant cells, these can be of different types such as cytoplasts, karyoplasts or mini-protoplasts or micro-protoplasts or microplasts.

- Cytoplasts lack the nucleus and contain the entire cytoplasm of a cell or part of it. These are often used for cybrid production.

- Karyoplasts or Mini-protoplasts contain a nucleus surrounded by some cytoplasm and the original outer plasma membrane.

- Micro-protoplasts contain fraction of both nuclear and cytoplasmic material.

- Microplasts lack the nuclear material and contain only fraction of cytoplasm and outer membrane.

Protoplast Viability Test

The most frequently used staining methods for checking protoplast viability are:

1. Fluorescein diacetate (FDA) dissolved in acetone is used at a conc. of 0.01% and intact viable protoplasts only fluoresce when observed under UV.

2. Phenosafranine is also used at a conc. of 0.01 %, which is specific for dead protoplast that shows red in colour.

Somatic Cell Hybridization

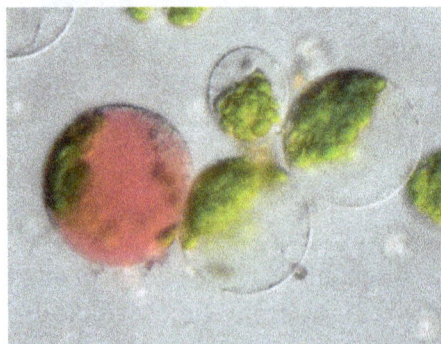

The process by which protoplasts of two different plant species fuse together to form hybrids is known as somatic hybridisation and the hybrids so produced is known as somatic hybrids.

Sexual hybridization since time immemorial has been used as a method for crop improvement but it has its own limitations as it can only be used within members of same species or closely related wild species. Thus, this limits the use of sexual hybridization as a means of producing better varieties. Development of viable cell hybrids by somatic hybridization, therefore, has been considered as an alternative approach for the production of superior hybrids overcoming the species barrier. The technique can facilitate breeding and gene transfer, bypassing problems associated with conventional sexual crossing such as, interspecific, intergeneric incompatibility. This technique of hybrid production via protoplast fusion allows combining somatic cells (whole or partial) from different cultivars, species or genera resulting in novel genetic combinations including symmetric somatic hybrids, asymmetric somatic hybrids or somatic cybrids.

The most common target using somatic hybridization is the gene of symmetric hybrids that contain the complete nuclear genomes along with cytoplasmic organelles of both parents. This is unlike sexual reproduction in which organelle genomes are generally contributed by the maternal parent. On the other hand, somatic cybridization is the process of combining the nuclear genome of one parent with the mitochondrial and/or chloroplast genome of a second parent. Cybrids can be produced by donor-recipient method or by cytoplast-protoplast fusion. Incomplete asymmetric somatic hybridization also provides opportunities for transfer of fragments of the nuclear genome, including one or more intact chromosomes from one parent (donor) into the intact genome of a second parent (recipient).

These methods provide genetic manipulation of plants overcoming hurdle of sexual incompatibility, thereby, serving as a method of bringing together beneficial traits from taxonomically distinct species which cannot be achieved by sexual crosses. Several parameters such as, source tissue, culture medium and environmental factors influence ability of a protoplast derived hybrid cells to develop into a fertile plant. The general steps involved in somatic hybridization and cybridization methods are elaborated in figures below.

Steps Involved in Somatic Hybridization

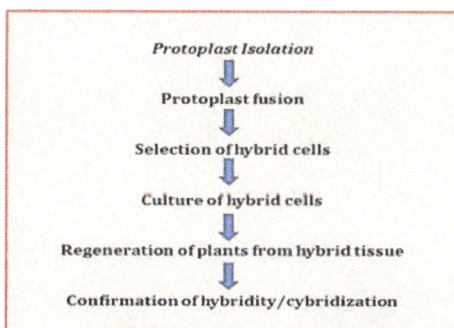

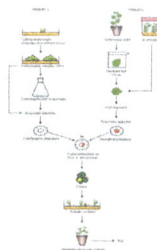

Protoplast Isolation

Protoplast fusion

Selection of hybrid cells

Culture of hybrid cells

Regeneration of plants from hybrid tissue

Confirmation of hybridity/cybridization

Figure: Schematic view of symmetric protoplast fusion producing somatic hybrids.

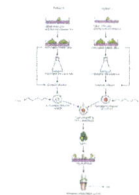

Figure: Schematic view of asymmetric protoplast fusion using donor-recipient method resulting into creation of alloplasmic somatic hybrid or cybrids.

Protoplast Fusion

Protoplast fusion could be spontaneous during isolation of protoplast or it can be induced by mechanical, chemical and physical means. During spontaneous process, the adjacent protoplasts fuse together as a result of enzymatic degradation of cell walls forming homokaryons or homokaryo-cytes, each with two to several nuclei. The occurrence of multinucleate fusion bodies is more frequent when the protoplasts are prepared from actively dividing callus cells or suspension cultures. Since the somatic hybridization or cybridization require fusion of protoplasts of different origin, the spontaneous fusion has no value. To achieve induced fusion, a suitable chemical agent (fusogen) like, $NaNO_3$, high Ca^{2+}, polyethylene glycol (PEG), or electric stimulus is needed.

i. Fusion by means of $NaNO_3$: It was first demonstrated by Kuster in 1909 that the hypotonic solution of $NaNO_3$ induces fusion of isolated protoplast forming heterokaryon (hybrid). This method was fully described by Evans and Cocking, however this method has a limitation of generating few no of hybrids, especially when highly vacuolated mesophyll protoplasts are involved.

ii. High pH and Ca^{++} treatment: This technique lead to the development of intra- and interspecific hybrids. It was demonstrated by Keller and Melcher in 1973. The isolated protoplasts from two plant species are incubated in 0.4 M mannitol solution containing high Ca^{++}(50 mM CaCl2.2H2O) with highly alkaline pH of 10.5 at 37°C for about 30 min. Aggregation of protoplasts takes place at once and fusion occurs within 10 min.

iii. Polyethylene glycol treatment: Polyethylene glycol (PEG) is the most popularly known fusogen due to ability of forming high frequency, binucleate heterokaryons with low cytotoxicity. With PEG the aggregation occurred mostly between two to three protoplasts unlike Ca^{++} induced fusion which involves large clump formation. The freshly isolated protoplasts from two selected parents are mixed in appropriate proportions and treated with 15-45% PEG (1500-6000MW) solution for 15-30 min followed by gradual washing of the protoplasts to remove PEG. Protoplast fusion occurs during washing. The washing medium may be alkaline (pH 9-10) and contain a high Ca^{++} ion concentration (50 mM). This combined approach of PEG and Ca^{++} is much more efficient than the either of the treatment alone. PEG is negatively charged and may bind to cation like Ca^{++}, which in turn, may bind to the negatively charged molecules present in plasma lemma, they can also bind to cationic molecules of plasma membrane. During the washing process, PEG molecules may pull out the plasma lemma components bound to them. This would disturb plamalemma organization and may lead to the fusion of protoplasts located close to each other. The technique is nonselective thus, induce fusion between any two or more protoplasts.

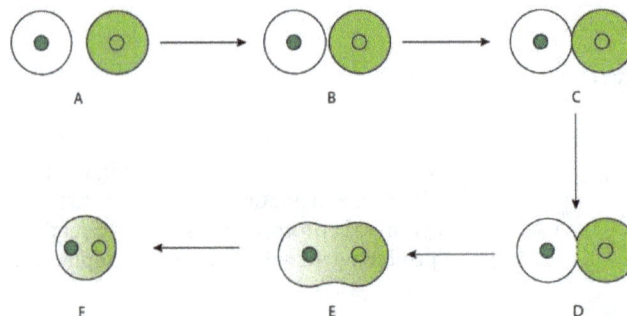

Figure: Sequential stages in protoplast fusion. (A) two separate protoplasts, (B) agglutination of two protoplasts, (C and D) Membrane fusion at localized site, and (E and F) development of spherical heterokaryon.

iv. Electrofusion: The chemical fusion of plant protoplast has many disadvantages – (1) The fusogen are toxic to some cell systems, (2) it produces random, multiple cell aggregates, and (3) must be removed before culture. Compare to this, electrofusion is rapid, simple, synchronous and more easily controlled. Moreover, the somatic hybrids produced by this method show much higher fertility than those produced by PEG-induced fusion.

Zimmermann and Scheurich demonstrated that batches of protoplasts could be fused by electric fields by devising a protocol which is now widely used. This protocol involves a two-step process. First, the protoplasts are introduced into a small fusion chamber containing parallel wires or plates which serve as electrodes. Second, a low-voltage and rapidly oscillating AC field is applied, which causes protoplasts to become aligned into chains of cells between electrodes. This creates complete cell-to-cell contact within a few minutes. Once alignment is complete, the fusion is induced by application of a brief spell of high-voltage DC pulses (0.125-1 kVcm-1). A high voltage DC pulses induces a reversible breakdown of the plasma membrane at the site of cell contact, leading to fusion and consequent membrane reorganization. The entire process can be completed within 15 min.

Selection of Fusion Products

The somatic hybridization by electrofusion of protoplasts allow one-to-one fusion of desired pairs of protoplasts and, therefore, it is easy to know the fate of fusion products. However, protoplast suspension recovered after chemical treatments (fusogen) consists of the following cell types:

- unfused protoplasts of the two species/strains,

- products of fusion between two or more protoplasts of the same species (homokaryons), and

- 'hybrid' protoplasts produced by fusion between one (or more) protoplasts of each of the two species (heterokaryons).

The heterokaryons which are the potential source of future hybrids constitute of a very small (0.5-10%) proportion of the mixture. Therefore, an effective strategy has to be employed for their identification and isolation. Various protocols have been proposed and practiced for the effective selection of hybrids, including morphological basis, complementation of biochemical and genetic traits of the fusing partners, and manual or electronic sorting of heterokaryons/hybrid cells.

Morpho-physiological basis: The whole mixtures of the protoplasts are cultured after fusion treatment and the resulting calli or regenerants are screened for their hybrid characteristics. Occasionally the hybrid calli outgrow the parental cell colonies and are identified by their intermediate morphology, i.e. green with purple coloured cells. However, the process is labour intensive and requires glasshouse facilities. It is limited to certain combinations showing differences in their regeneration potential under specific culture conditions.

Complementation: In this case complementation or genetic or metabolic deficiencies of the two fusion partners are utilized to select the hybrid component. When protoplasts of two parents, (one parent bearing cytoplasmic albino trait and the other parent bearing green trait) each parent carrying a non-allelic genetic or metabolic defect are fused, it reconstitutes a viable hybrid cell, of wild type in which both defects are mutually abolished by complementation, and the hybrid cells are able to grow on minimal medium non-permissive to the growth of the parental cells bearing green trait. Later, the calli of hybrid nature could be easily distinguished from the parental type tissue (albino trait) by their green color. The complementation selection can also be applied to dominant characters, such as dominant resistance to antibiotics, herbicides or amino acid analogues.

Isolation of heterokaryons or hybrid cells: The manual or electronic isolation of heterokaryons or hybrid cells is the most reliable method. Manual isolation requires that the two parental type protoplasts have distinct morphological markers and are easily distinguishable. For example, green vacuolated, mesophyll protoplasts from one parent and richly cytoplasmic, non green protoplasts from cultured cells of another parent. The dual fluorescence method also helps easy identification of fusion products. In this case, the protoplast labeled green by treatment with fluorescein diacetate (FDA, 1-20 mgl-1) are fused with protoplasts emitting a red fluorescence, either from chlorophyll autofluorescence or from exogenously applied rhodamine isothiocyanate (10-20 mgl-1). The labeling can be achieved by adding the compound into the enzyme mixture. This can be applied even for morphologically indistinguishable protoplasts from two parents.

Verification and Characterization of Somatic Hybrids

As no system is foolproof and they have their own advantages and disadvantages. Therefore, even after selecting the desired hybrids/cybrids following protoplast fusion, it is required to carry out one or more tests to compare the parent protoplast lines with the putative hybrids. Some of the techniques that can be tried are:

Morphology: Somatic hybrids in most of the cases show characters intermediate between the two parents such as, shape of leaves, pigmentation of corolla, plant height, root morphology and other vegetative and floral characters. The method is not much accurate as tissue culture conditions may also alter some morphological characters or the hybrid may show entirely new traits not shown by any of the parents.

Isozyme analysis: Multiple molecular forms of same enzyme which catalyses similar or identical reactions are known as isozymes. Electrophoresis is performed to study banding pattern as a check for hybridity. If the two parents exhibit different band patterns for a specific isozyme the putative hybrid can be easily verified. The isozymes commonly used for hybrid identification include, acid phosphatase, esterase, peroxidase.

Cytological analysis: Chromosome counting of the hybrid is an easier and reliable method to ensure hybridity as it also provides the information of ploidy level. Cytologically the chromosome count of the hybrid should be sum of number of chromosomes from both the parents. Besides number of chromosomes, the size and structure of chromosomes can also be monitored. However, the approach is not applicable to all species, particularly where fusion involves closely related species or where the chromosomes are very small. Moreover, sometimes the somaclonal variations may also give rise to different chromosome number.

Molecular analysis: Specific restriction pattern of nuclear, mitochondrial and chloroplast DNA characterizes the plastomes of hybrids and cybrids. Molecular markers such as RFLP, RAPD, ISSR can be employed to detect variation and similarity in banding pattern of fused protoplasts to verify hybrid and cybrid.

Cybrids or Cytoplasmic Hybrids

Sexual hybridization involves fusion of the nuclear genes of both the parents but somatic hybrids involves even cytoplasm from both the parental species in hybrid obtained by protoplast fusion. However, in another case somatic hybrids containing nuclear genome of one parent but cytoplasm from both the parents, are termed as cybrids. The approach is time consuming and require several years of crossing plants provides an opportunity to study interparental mitochondrial, chloroplast fusion giving rise to plants with novel genomes.

Methods to Produce Cybrids

They are produced in variable frequencies in normal protoplast fusion experiments due to one of the following methods:

1. Fusion of normal protoplast with an enucleated protoplast. The enucleated protoplast can be produced by high speed centrifugation (20,000-40,000xg) for 60 min with 5-50% percoll.

2. Fusion between a normal protoplast and another protoplast with a non-viable nucleus or suppressed nucleus.

3. Elimination of one of the nuclei after heterokaryons formation.

4. Selective elimination of chromosomes at a later stage.

5. Irradiating (with X-rays or gamma rays) the protoplasts of one species prior to fusion in order to inactivate their nuclei.

6. By preparing enucleate protoplasts (cytoplasts) of one species and fusing them with normal protoplasts of the other species.

Cybrids provide the following unique opportunities: (i) transfer of plasmogenes of one species into the nuclear background of another species in a single generation, and even in (ii) sexually incompatible combinations, (iii) recovery of recombinants between the parental mitochondrial or chloroplast DNAs (genomes), and (iv) production of a wide variety of combinations of the parental and recombinant chloroplasts with the parental or recombinant mitochondria.

Applications of Somatic Hybridization

1. Novel interspecific and intergeneric crosses which are difficult to produce by conventional methods can be easily obtained.

2. Important characters, such as resistance to diseases, ability to undergo abiotic stress and other quality characters, can be obtained in hybrid plant by the fusion of protoplasts of plant bearing particular character to the other plant which may be susceptible to diseases.

3. Protoplasts of sexually sterile haploid, triploid, aneuploid plants can be fused to obtain fertile diploids and polyploids.

4. Studying cytoplasmic genes may be helpful to carry out plant breeding.

5. Most of the agronomically important traits, such as cytoplasmic male sterility, antibiotic resistance and herbicide resistance, are cytoplasmically encoded, hence can be easily transferred to other plant.

6. Plants in juvenile stage can also be hybridized by means of somatic hybridization.

7. Somatic hybridization can be used as a method for the production of autotetraploids.

Limitations of Somatic Hybridization

1. Application of protoplast methodology requires efficient plant regeneration system from isolated protoplasts. Protoplasts from two species can be fused, however, production of somatic hybrids is not easy.

2. Lack of a proper selection method for fused products (hybrids) poses a problem.

3. The end product of somatic hybridization are often unbalanced (sterile, misformed and unstable).

4. Somatic hybridization of two diploids leads to formation of amphidiploids which is unfavorable.

5. It is not sure for a character to completely express after somatic hybridization.

6. The regeneration products of somatic hybridization are often variable due to somaclonal variation, chromosome elimination, organelle segregation.

7. All diverse intergeneric somatic hybrids are sterile and, therefore, have limited chances of development of new varieties.

8. To transfer useful genes from wild species to cultivated crop, it is necessary to achieve intergeneric recombination or chromosome substitution between parental genomes.

Gene Gun Method

The gene gun method is a method used for genetically modifying plants.

With the use of a gene gun, the gene gun method delivers extra DNA directly into a plant's nucleus. The method is also commonly called particle acceleration or microprojectile bombardment.

The gene gun can be used on seedlings or tissue culture cells. Prior to injecting the DNA into the plant tissue via the gene gun, either microscopic gold or tungsten particles are liberally coated with many hundreds of copies of genes. The particles are then forced into the nucleus with the gene gun.

The first gene gun method of DNA modification was developed in 1980. The process relied on a 22-caliber cartridge to shoot the gene-coated metal particles into the plant's cell.

Nowadays, modern versions of the gene gun operate differently. The modernized gene gun depends on a vacuum chamber and high-pressure gas to hurl the metal particles into the plant cells.

With each firing of the gene gun, the pressurized gas is released in a sudden popping burst and the gene coated metal particles fire into the cells.

The gene gun is highly effective at modifying the DNA and genetics of plant cells.

Genetic modification in plants can make them resistant to drought, disease, and pests. It also helps create a more nutritious edible plant with higher levels of proteins.

This method was used to create several commercial events such as RoundupReady soybean, it has several limitations that make it less appealing than the use of Agrobacterium as a transformation method.

The gene gun is a good example of a creative idea being developed into a practical technology.

The gene gun can be used on either tissue culture cells or seedlings (to make chimeric plants) of any species. As the name implies, this method works by shooting DNA into the plant cells.

Gene Gun Design

The gene gun was originally a Crosman air pistol modified to fire dense tungsten particles. It was invented by John C Sanford, Ed Wolf and Nelson Allen at Cornell University, and Ted Klein of DuPont, between 1983 and 1986. The original target was onions (chosen for their large cell size) and it was used to deliver particles coated with a marker gene. Genetic transformation was then proven when the onion tissue expressed the gene.

The earliest custom manufactured gene guns (fabricated by Nelson Allen) used a 22 caliber nail gun cartridge to propel an extruded polyethylene cylinder (bullet) down a 22 cal. Douglas barrel. A droplet of the tungsten powder and genetic material was placed on the bullet and shot down the barrel at a lexan"stopping" disk with a petri dish below. The bullet welded to the disk and the genetic material blasted into the sample in the dish with a doughnut effect (devastation in the middle, a ring of good transformation and little around the edge). The gun was connected to a vacuum pump and was under vacuum while firing. The early design was put into limited production by a Rumsey-Loomis (a local machine shop then at Mecklenburg Rd in Ithaca, NY, USA). Later the design was refined by removing the "surge tank" and changing to nonexplosive propellants. DuPont added a plastic extrusion to the exterior to visually improve the machine for mass production to the scientific community. Bioradcontracted with Dupont to manufacture and distribute the device. Improvements include the use of helium propellant and a multi-disk-collision delivery mechanism. Other heavy metals such as gold and silver are also used. Gold may be favored because it has better uniformity than tungsten and tungsten can be toxic to cells, but its use may be limited due to availability and cost.

Biolistic Construct Design

A construct is a piece of DNA inserted into the target's genome, including parts that are intended to be removed later. All biolistic transformations require a construct to proceed and while there is great variation among biolistic constructs, they can be broadly sorted into two categories: those which are designed to transform eukaryotic nuclei, and those designed to transform prokaryotic-type genomes such as mitochondria, plasmids or plastids.

Those meant to transform prokaryotic genomes generally have the gene or genes of interest, at least one promoter and terminator sequence, and a reporter gene; which is a gene used to enable detection or removal of those cells which didn't integrate the construct into their DNA. These genes may each have their own promoter and terminator, or be grouped to produce multiple gene products from one transcript, in which case binding sites for translational machinery should be placed between each to ensure maximum translational efficiency. In any case the entire construct is flanked by regions called border sequences which are similar in sequence to locations within the genome, this allows the construct to target itself to a specific point in the existing genome.

Constructs meant for integration into a eukaryotic nucleus follow a similar pattern except that: the construct contains no border sequences because the sequence rearrangement that prokaryotic constructs rely on rarely occurs in eukaryotes; and each gene contained within the construct must be expressed by its own copy of a promoter and terminator sequence.

Though the above designs are generally followed, there are exceptions. For example, the construct might include a Cre-Lox system to selectively remove inserted genes; or a prokaryotic construct may insert itself downstream of a promoter, allowing the inserted genes to be governed by a promoter already in place and eliminating the need for one to be included in the construct.

Application

Gene guns are so far mostly used with plant cells. However, there is much potential use in humans and other animals as well.

Plants

The target of a gene gun is often a callus of undifferentiated plant cells or a group of immature embryos growing on gel medium in a Petri dish. After the DNA-coated gold particles have been delivered to the cells, the DNA is used as a template for transcription (transient expression) and sometimes it integrates into a plant chromosome ('stable' transformation).

If the delivered DNA construct contains a selectable marker, then stably transformed cells can be selected and cultured using tissue culture methods. For example, if the delivered DNA construct contains a gene that confers resistance to an antibiotic or herbicide, then stably transformed cells may be selected by including that antibiotic or herbicide in the tissue culture media.

Transformed cells can be treated with a series of plant hormones, such as auxins and gibberellins, and each may divide and differentiate into the organized, specialized, tissue cells of an entire plant. This capability of total re-generation is called totipotency. The new plant that originated from a successfully transformed cell may have new traits that are heritable. The use of the gene gun may be contrasted with the use of Agrobacterium tumefaciens and its Ti plasmid to insert DNA into plant cells.

Advantages

Biolistics has proven to be a versatile method of genetic modification and it is generally preferred to engineer transformation-resistant crops, such as cereals. Notably, Bt maize is a product of biolistics. Plastid transformation has also seen great success with particle bombardment when compared to other current techniques, such as Agrobacterium mediated transformation, which have difficulty targeting the vector to and stably expressing in the chloroplast. In addition, there are no reports of a chloroplast silencing a transgene inserted with a gene gun. Additionally, with only one firing of a gene gun, a skilled technician can generate two transformed organisms. This technology has even allowed for modification of specific tissues in situ, although this is likely to damage large numbers of cells and transform only some, rather than all, cells of the tissue.

Limitations

Biolistics introduces DNA randomly into the target cells. Thus the DNA may be transformed into whatever genomes are present in the cell, be they nuclear, mitochondrial, plasmid or any others, in any combination, though proper construct design may mitigate this. Another issue is that the gene inserted may be overexpressed when the construct is inserted multiple times in either the same or different locations of the genome. This is due to the ability of the constructs to give and take genetic material from other constructs, causing some to carry no transgene and others to carry multiple copies; the number of copies inserted depends on both how many copies of the transgene an inserted construct has, and how many were inserted. Also, because eukaryotic constructs rely on illegitimate recombination, a process by which the transgene is integrated into the genome without similar genetic sequences, and not homologous recombination, which inserts at similar sequences, they cannot be targeted to specific locations within the genome, unless the transgene is co-delivered with genome editing reagents.

Electroporation

Electroporation is a popular physical method for introducing new genes directly into the protoplasts.

In this method, electric field is playing important role. Due to the electric field protoplast get temporarily permeable to DNA. In electroporation, plant cell protoplasts are kept in an ionic solution containing the vector DNA in a small chamber that has electrodes at opposite ends. A pulse of high voltage is applied to the electrode which makes the transient pores (ca. 30 nm) in the plasma membrane, allowing the DNA to diffuse into the cell. Immediately, the membrane reseals.

If appropriately treated, the cells can regenerate cell wall, divide to form callus and, finally, re-generate complete plants in suitable medium. The critical part of the procedure is to determine conditions which produce pores that are sufficiently large and remain open long enough to allow for DNA diffusion. At the same time, the conditions should make pores that are temporary. With a 1 cm gap between the electrodes and protoplasts of 40-44μm diameter, 1-1.5 kVcm -2 of field strength for 10μs is required for efficient introduction of DNA. It was seen that presence of 13% PEG (added after DNA) during electroporation significantly raised the transformation frequency. The other factors which may improve the transformation frequency by electroporation are linear-izing of plasmid, use of carrier DNA, and heat shock (45 ~ for 5 min) prior to addition of vector, and placing on ice after pulsing. Under optimal conditions transformation frequencies of up to 2% have been reported. Stably transformed cell lines and full plants of a number of cereals have been produced through electroporation.

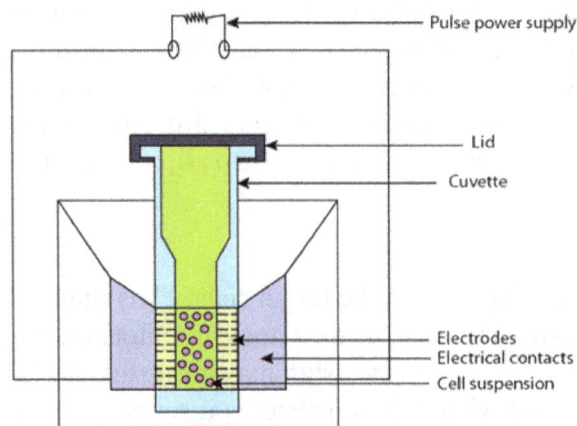

Figure: Electroporation

There are some parameters that can be considered when performing in vitro electroporation:

Cell size

Cell size is inversely correlated to the size of the external field needed to generate permeabilization. Consequently, optimization for each cell type is essential. Likewise, cell orientation matters for cells that are not spherical.

Temperature

It has been observed that plant membrane resealing is effectively temperature dependent and shows slow closure at low temperatures. For DNA transfer, it has been found that cooling at the time of permeabilization and subsequent heating in incubator increases transfer efficacy and cell viability.

Post-pulse Manipulation

Cells are susceptible when in the permeabilized state, and it has been shown that waiting for 15min after electroporation in order to allow resealing before pipetting cells, increases cell viability.

Composition of Electrodes and Pulsing Medium

For short pulses is needed for release of metal from the standard aluminium electrodes used in standard disposable cuvettes. Some authors advocate the use of low conductivity or more resistance media for DNA transfer in order to increase viability and increase transfection efficacy.

Microinjection

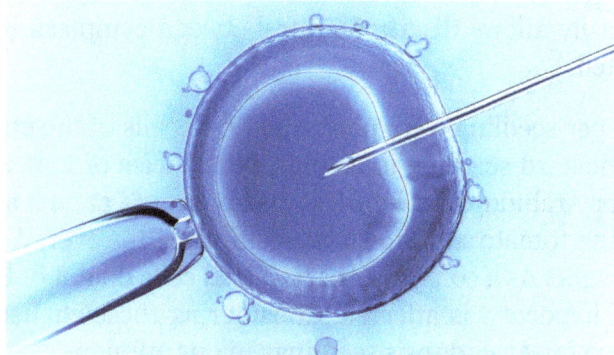

Microinjection is an established method to introduce DNA into single cells to generate transient and stable transformants. The injection of single cells within plant organs, however, is more difficult, and only a low percentage of the injected cells survive. The Eppendorf micromanipulation system has been used to inject tomato, mustard and Arabidopsis seedlings.

Usually plant cells possess a very rigid cell wall, which is a problem for microinjection since it requires stable injection capillaries with small diameters. In combination with the cellular turgor, the wall is responsible for the structural integrity of a plant seedling. The cell turgor of plants can be very high and is not only responsible for the morphology of a seedling but also plays an important role in development. Even if the injection against the high internal pressure was successful, the injury of a larger cell wall area will automatically lead to the destruction of the cell, at the latest when the needle is retreated.

The large central vacuole of a plant cell is also a problem that diminishes the chances of a successful cytoplasmic injection. The cytoplasm surrounding the central vacuole is only a few micrometres thick. As a result of the deformation of the cell wall during the injection and the sudden relaxation when the capillary penetrates the wall, the needle will often end up in the vacuole. Material injected into the vacuole is not only trapped but in many cases degraded, due to the low pH and the activity of several proteases. In addition, the tonoplast — the single membrane surrounding the vacuole — is easily injured, which leads to leakage of cytotoxic substances into the cytoplasm and hence to cell death.

Furthermore, because of multiple cell layers, light scattering and pigmentation, the optical properties of whole plant seedlings or plant organs are not optimal. Therefore, it is often difficult to identify different cell layers or even subcellular structures like vacuoles under microscopic control during the microinjection process.

To approach the described problems, Eppendorf developed a micromanipulation system equipped with very fast micromotors, variable injection parameters, high holding and injection pressure, in combination with a piezo step function. The set-up was tested for its suitability to inject plant cells — specifically tomato, mustard and Arabidopsis seedlings.

The micromanipulation system's injection angle was adjusted to 35° and axial injection movements were performed at a speed of 3000 μm/s with a step size of 10–20 μm. These settings ensure minimal injuries of the cell wall and thus raise the survival rate after injection. The injection movement can be supported by synchronised piezo impulses when the system's InjectMan 4 and PiezoXpert devices are electronically connected. As a marker for injection, Lucifer yellow (ly) — a water-soluble dye that easily allows discrimination between cytoplasmic and vacuolar localised fluorescence — was injected.

Typically, 5–8 injections per seedling were performed into cells of the etiolated hypocotyl. In the case of the tomato and mustard seedling injections, efficiencies of 26% and 48% were obtained. The injection efficiency for Arabidopsis seedlings was about 10% at 24 h after injection — strongly decreased compared to the tomato and mustard seedlings. However, Arabidopsis seedlings are much smaller and very fragile. As a consequence, injected cells do not only die relatively often but in many cases the whole hypocotyl is affected. Considering these challenges, 10% injection efficiency reflects a good value for Arabidopsis seedlings in our opinion.

Localised fluorescence of ly 24 h after injection indicated that the cells were viable, but ly is no in vivo marker. Hence, we decided to inject Alexa Fluor 488-labelled histone (histone-AF488). After injection into the cytoplasm, the histones should be transported into the nucleus, where they accumulate. Histone import into the nucleus is an active process that requires GTP and intact protein complexes. Therefore, the import of histone-AF488 should only occur in living cells. The percentage of cells showing histone-AF488 fluorescence was in the same range as the observed percentage after ly injection, confirming that these cells survived the injection procedure.

The Eppendorf micromanipulation system, with its fast and precise motor movements, was thus well suited for the injection of plant cells. This includes the injection of Arabidopsis seedlings with relatively high efficiencies. Since a large collection of Arabidopsis mutants and transgenic lines is available for many research areas, microinjection of Arabidopsis cells from specific organs and tissues should, in addition to injection of other plant species, help to analyse numerous scientific questions.

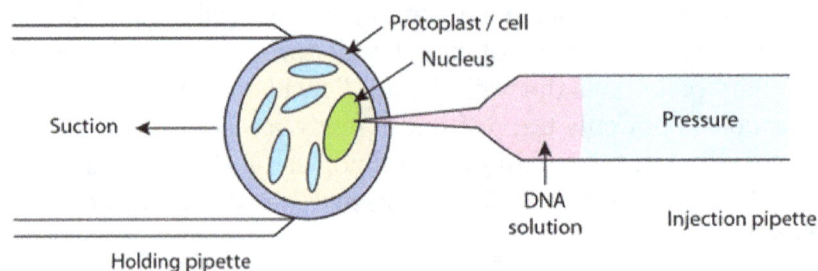

Agrobacterium

Agrobacterium elicits neoplastic growths (called crown gall tumors) that affect most dicotyledonous plants. Moreover, although plants represent the natural hosts for Agrobacterium, this microorganism can also genetically transform a wide range of other eukaryotic species, from yeast to mushrooms and filamentous fungi to phytopathogenic fungi to human cells. Most functions for Agrobacterium-host cell DNA transfer are coded by a large (200-kb) tumor-inducing (Ti) plasmid that resides in the bacterial cell and carries two important genetic components: the vir (virulence) region and the T-DNA delimited by two 25-bp direct repeats at its ends, termed the T-DNA borders. The vir region comprises seven major loci, virA, virB, virC, virD, virE, virG, and virH, which encode most of the bacterial protein machinery (Vir proteins) of the DNA transport. After induction of vir gene expression by small phenolic signal molecules secreted from wounded susceptible plant cells, the T-DNA borders are nicked by the bacterial VirD2 endonuclease, generating a transferable single-stranded (ss) copy of the bottom strand of the T-DNA region, designated the T strand.

Interestingly, the T strand does not travel alone but is thought to directly associate with two Agrobacterium proteins, VirD2 and VirE2, forming a transport (T) complex in which one molecule of VirD2 is covalently attached to the 5′-end of the T strand, whereas VirE2, an ssDNA-binding protein, is presumed to cooperatively coat the rest of the T strand molecule. Although only the wild-type T-DNA carries Ti genes, any DNA placed between the T-DNA borders will be transferred to the plant host. This lack of sequence specificity implies that a T-DNA molecule itself does not encode protein machinery for its transport from the bacterial cell into the host cell, import into the host cell nucleus, and integration into the host cell genome. Instead, these functions are fulfilled by the bacterial Vir proteins and their host cell partners.

In the last quarter of a century, since the discovery of the stable integration of the bacterial DNA in crown gall tumors, Agrobacterium has served as a primary a tool for plant genetic engineering. Furthermore, the Agrobacterium-host cell interaction also represents a unique and powerful experimental system to study a wide spectrum of basic biological processes such as cell-cell recognition and cell-to-cell transport, nuclear import, assembly and disassembly of protein-DNA complexes, DNA recombination, and regulation of gene expression.

Genetic Engineering

Predating the science fiction visions of Nano machines performing genetic engineering and other biotechnological tasks, Agrobacterium is a present-day microscopic but extremely complex machine routinely used to alter genotypes of higher plants. This use of Agrobacterium is based on its unique capacity for "trans-kingdom sex", i.e. transfer of genetic material between prokaryotic and eukaryotic cells. Decades of research altered, augmented, and vastly improved this natural capacity, resulting in ingeniously modified Agrobacterium strains that can transfer and stably integrate virtually any gene to a variety of plant species, from research model plants such as Arabidopsis to agriculturally important rice and corn. In this issue, Valentine reviews the molecular mechanisms of T-DNA transfer that underlie the use of Agrobacterium as gene vector. Importantly, a special emphasis is made on the ethical and political aspects of the release of Agrobacterium-generated genetically modified plants into the environment and on the attitudes of different and often opposing geopolitical and economic forces to these matters.

Cell-Cell Recognition and Attachment

Agrobacterium recognition of and attachment to the host cells is an early and essential step of the infection process. The bacterial proteins participating in these events are encoded by several loci, e.g. chvA, chvB, pscA, and att. In particular, the att genes, such as attR and attD, are located on a cryptic plasmid pAtC58. Although pAtC58 is about twice as large as the Ti plasmid and, thus, likely encodes numerous bacterial functions, it has been considered dispensable for infection. In this issue, Nair et al. revisit this dogma and demonstrate that pAtC58 increases Agrobacterium virulence. Surprisingly, however, this increase was likely due to the enhancement of the vir gene expression rather than to the presence of the attR gene on the pATC58 plasmid.

Although the bacterial genes participating in Agrobacterium attachment to the host plants are relatively well characterized, the involvement of host factors remains largely obscure. Recent progress in this direction stemmed from identification of T-DNA-tagged Arabidopsis mutants defective in their ability to bind Agrobacterium. In this issue, Zhu et al. (2003) report characterization of one such mutant, rat4, which contains a T-DNA insertion in the 3'-untranslated region of cellulose synthase-like gene CSLA9. Their data indicate that CSLA9 is involved in development and growth of lateral roots, determination of sugar composition of plant cell walls, and the ability of the roots to bind Agrobacterium. Furthermore, the CSLA9 promoter exhibited enhanced expression in the root elongation, previously shown to be most susceptible to Agrobacterium-mediated transformation. Similarly, preferential expression in the elongation zones of Arabidopsis roots has been reported for a plant histone gene, H2A-1, which is required for T-DNA integration.

Cell-to-Cell Transport

T strands with their cognate VirD2 are exported into the host cell by a type IV secretion system, which in Agrobacterium is assembled from proteins encoded by the virD4 gene and virB operon, with 11 open reading frames. Interestingly, this system also exports other Agrobacterium proteins, such as VirE3, VirF, and VirE2. Although VirE2 most likely packages the T strand into the T complex, this binding is thought to occur within the cytoplasm of the host cell after independent export of the T strands and VirE2. But what prevents VirE2 from binding to the T strands already within

Agrobacterium? This role has been assigned to the VirE1 chaperone protein that associates with VirE2 and blocks its binding to ssDNA. Furthermore, VirE1 has been suggested also to participate in the VirE2 export. However, convincing evidence to the contrary is presented by Vergunst. Using the Cre Recombinase Reporter Assay for Translocation to study protein export from Agrobacterium into plant and yeast cells, they demonstrate that recognition of VirE2 by the bacterial export machinery and its subsequent translocation into host cells does not depend on the presence of VirE1. Cre Recombinase Reporter Assay for Translocation was then utilized to show that VirE3, another product of the virE locus earlier reported to be exported into the yeast cells, is transferred from Agrobacterium to plants. The function of VirE3 in the plant cell, however, remains completely unknown. Vergunst also localized the Agrobacterium-to-plant export signal to the C-terminal 50 amino acids of VirE2 and VirE3. These and previous studies clearly demonstrate that in addition to DNA, Agrobacterium transfers a variety of its own proteins to the host cell; the challenge to researchers now is to understand how these bacterial proteins participate in the genetic transformation process from within the host cell.

Dna Integration and Expression

T-DNA integration is the culmination point of the entire process of the Agrobacterium-plant cell DNA transfer. But how does a T-DNA molecule insinuate itself into the molecule of the plant genomic DNA? Because the T-DNA does not encode enzymatic activities required for integration, the protein components of the T-complex, i.e. VirD2 and VirE2 and plant DNA ligases, must provide these functions. T-DNA integration was proposed to initiate with ligation of the 5' end of the T strand to the genomic DNA followed by second strand synthesis by the plant DNA repair machinery. However, another study suggested that the T strand is converted into a double-stranded form before integration.

Obviously, it is the expression of the integrated transgenes that produces tumors in the wild-type infection or desired transgenic phenotypes in genetic engineering experiments. However, some transgenes, although stably integrated in the plant genome, often are not expressed due to their posttranscriptional gene silencing (PTGS), which is characterized by a reduction in transcript levels without affecting the rate of transcription. Furthermore, PTGS of a transgene can also silence its endogenous cellular homologs, resulting in co-suppression. Agrobacterium-mediated gene delivery has served as a valuable experimental tool to study the mechanisms of induction and cell-to-cell spread of PTGS. Continuing this trend, Lee utilized silencing of Agrobacterium oncogenes contained within the T-DNA to demonstrate that transgene sequences influence the effectiveness of PTGS and that sequences required for oncogene silencing must include a translation start site. Unexpectedly, unlike several other cases of PTGS, oncogene silencing was not graft transmissible. In addition to helping us better understand the mechanisms of PTGS, silencing of Agrobacterium oncogenes represents a novel approach to control the crown gall disease, which affects such agronomically important plants as grape, rose, apple, cherry, and others.

Finally, Agrobacterium infection was also used as a model system to study cellular processes required for establishment of plant tumors. Wächter showed that vascular differentiation and disruption of epidermal cell layers play key roles in tumor formation, allowing delivery of water and Suc to the proliferating tumor cells. This work suggested that Agrobacterium-induced tumors represent nutritional sinks to which a high-volume flow of solutes with essential inorganic nutrients is directed in an ethylene-dependent fashion, followed by a Suc-to-hexose shift in sugar balance caused by vacuolar invertases within the growing tumors.

Examples of Genetic Modification

Bananas

In many countries around the world bananas are the main source of calories. According to reports from Uganda, their production is compromised by the emergence of new diseases. Ugandan scientists have successfully used a genetic modification, inserting a pepper gene into bananas, which prevents the fruit from getting the disease.

Onions that do not Make you Cry

In 2008, a New Zealand research team lead by Colin Eady produced an onion that does not make you cry while cutting it. Interestingly, the insertion of a single gene which down regulates the activity of the onion enzyme that makes your eyes water has managed to achieve two things: firstly, onions no longer make your eyes water, and secondly, they now have even more health beneficial sulphur-containing substances than regular onions.

Golden Rice

On 31st July 2000, Ingo Potrykus appeared on the cover of Time magazine. The Swiss scientist and his German colleague Peter Beyerhad had produced a breed of rice which, unlike any other, also contains provitamin A. The lack of this vitamin is especially harmful to the poorest and is estimated to cause blindness among 250,000-500,000 children every year. Another two million people a year die from other deficiency-related causes. So far, the measures taken to introduce vitamin supplements have not yet reached those poor countries. Because of its colour, the product was first given the name golden rice, which remains the same to this day. After concluding numerous tests, researchers from the International Rice Research Institute in the Philippines have proven that even small amounts of the rice are sufficient, and that it is absolutely safe. Unfortunately, partly due to the vandalism of "green" activists, the rice has yet to reach its target group. We should also mention the existence of genetically modified rice, produced in 2011, which contains four times as much iron as the regular one, and could therefore save even more lives.

Purple Tomatoes

In 2008, a small English research group published a study describing how they had transferred a gene from a decorative plant into a tomato, which enabled the production of anthocyanin, making the tomato dark blue. They later tested it on mice, and discovered that it prevents them from getting cancer. These tomatoes cannot be bought due to their GMO status but you can buy similar dark blue ones, which are the result of a complex interspecific hybridisation performed by Italian scientists. Producing such tomatoes seems to have sparked competition between cultivators. In 2012 for instance, the Israelis announced that they have beaten the Italians by introducing a new species to the market called Black Galaxy.

Carrots that Help Prevent Osteoporosis

In 2004, an American research team transferred a CAX1 gene from the mouseear cress into carrots so that they contained larger amounts of organically bound calcium. In 2008, they performed a study where such carrots were tested on mice and 30 volunteers, and the results showed that humans absorbed 42% more calcium from the modified carrots than from regular ones. The aim of this test was to help prevent osteoporosis, while the emphasis was on its bioavailability in target tissues.

Soybean Oil for Frying

Two American companies have significantly improved soybean edible oil through genetic modification. The Plenish oil marketed by DuPont used gene silencing to produce oil that contains low levels of polyunsaturated fats and high levels of monounsaturated fats, while the saturated fatty acids have decreased by 20%. This kind of oil has high stability during baking, meaning it does not need to go through the chemical process of hydrogenation, which produces unwanted trans fats. There is a similar vascular-system-friendly oil marketed by Monsanto with two inserted genes enabling the production of omega-3 fatty acids. Rapeseed is also close to being released into production by the company BASF, containing five genes isolated from seaweed. These are the first products since the Flavr Savr tomato in 1994 to be made directly available to the consumer.

Arctic Apple

An apple turns brown if it is cut in half, which is why its slices are often soaked in antioxidants to prevent this. In 2012, Canadian scientists started releasing two popular apple cultivars, Golden Delicious and Granny Smith, and are planning to introduce Gala and Fuji cultivars, all under the brand name Arctic Apple. The non-browning characteristic could already be found in some apple cultivars, like the newest Slovene cultivar called Majda, but none of them has gained strong commercial recognition. The aim of this research is to prevent the most popular cultivars from turning brown. The same principle applies to these apples as to the previously mentioned onions: it is not about adding genes but about downregulating the activity of the existing ones.

References

- Duncan, R. (1996). "Tissue Culture-Induced Variation and Crop Improvement". Advances in Agronomy. 58: 201–40. doi:10.1016/S0065-2113(08)60256-4. ISBN 9780120007585.

- Transgenic-plants-meaning-reasons-and-fundamentals-588: biotechnologynotes.com, Retrieved 12 March 2018

- Bruening, G.; Lyons, J. M. (2000). "The case of the FLAVR SAVR tomato". California Agriculture. 54 (4): 6–7. doi:10.3733/ca.v054n04p6.

- Cisgenics-and-its-Application-in-Crop-Improvement-3861: biotecharticles.com, Retrieved 20 April 2018

- Andrew Pollack (19 June 2013). "Executive at Monsanto wins global food honor". The New York Times. Retrieved 20 June 2013.

- Gene-gun-method-3145: maximumyield.com, Retrieved 28 May 2018

- Koornneef, M.; Meinke, D. (2010). "The development of Arabidopsis as a model plant". The Plant Journal. 61 (6): 909–21. doi:10.1111/j.1365-313X.2009.04086.x. PMID 20409266.

- Microinjection-into-plant-cells-of-etiolated-seedlings-981244058: labonline.com.au, Retrieved 30 May 2018

- G., Halford, Nigel (2012). Genetically modified crops. World Scientific (Firm) (2nd ed.). London: Imperial College Press. ISBN 978-1848168381. OCLC 785724094.

- 10-successful-examples-of-genetic-modification: metinalista.si, Retrieved 25 June 2018

- Predieri, S. (2001). "Mutation induction and tissue culture in improving fruits". Plant Cell, Tissue and Organ Culture. 64 (2/3): 185–210. doi:10.1023/A:1010623203554.

Chapter 6

Microtechniques in Plant Biotechnology

Various microtechniques are used in plant biotechnology for the identification and characterization of plant species, which have significant implications in quality control and various other uses. This chapter carefully analyzes the varied microtechniques used in plant biotechnology, such as microscopy, cell sorting, plant histological technology, etc.

Microscopy

A microscope is an instrument that magnifies an object. Most photographs of cells are taken with a microscope, and these images can also be called micrographs.

The optics of a microscope's lenses change the orientation of the image that the user sees. A specimen that is right-side up and facing right on the microscope slide will appear upside-down and facing left when viewed through a microscope, and vice versa. Similarly, if the slide is moved left while looking through the microscope, it will appear to move right, and if moved down, it will seem to move up. This occurs because microscopes use two sets of lenses to magnify the image. Because of the manner by which light travels through the lenses, this system of two lenses produces an inverted image (binocular, or dissecting microscopes, work in a similar manner, but include an additional magnification system that makes the final image appear to be upright).

Light Microscopes

To give you a sense of cell size, a typical human red blood cell is about eight millionths of a meter or eight micrometers (abbreviated as eight μ m) in diameter; the head of a pin of is about two thousandths of a meter (two mm) in diameter. That means about 250 red blood cells could fit on the head of a pin.

Most student microscopes are classified as light microscopess. Visible light passes and is bent through the lens system to enable the user to see the specimen. Light microscopes are advantageous for viewing living organisms, but since individual cells are generally transparent, their components are not distinguishable unless they are colored with special stains. Staining, however, usually kills the cells.

Light microscopes commonly used in the undergraduate college laboratory magnify up to approximately 400 times. Two parameters that are important in microscopy are magnification and resolving power. Magnification is the process of enlarging an object in appearance. Resolving power is the ability of a microscope to distinguish two adjacent structures as separate: the higher the resolution, the better the clarity and detail of the image. When oil immersion lenses are used for the study of small objects, magnification is usually increased to 1,000 times. In order to gain a better understanding of cellular structure and function, scientists typically use electron microscopes.

An electron microscope is a type of microscope that uses electrons to illuminate a specimen and create an enlarged image. Electron microscopes have much greater resolving power than light microscopes and can obtain much higher magnifications. Some electron microscopes can magnify specimens up to 2 million times, while the best light microscopes are limited to magnifications of 2000 times. Both electron and light microscopes have resolution limitations, imposed by their wavelength. The greater resolution and magnification of the electron microscope is due to the wavelength of an electron, its de Broglie wavelength, being much smaller than that of a light photon, electromagnetic radiation.

An Electron Microscope.

The electron microscope uses electrostatic and electromagnetic lenses in forming the image by controlling the electron beam to focus it at a specific plane relative to the specimen in a manner similar to how a light microscope uses glass lenses to focus light on or through a specimen to form an image.

The first electron microscope prototype was built in 1931 by German engineers Ernst Ruska and Max Knoll. Although this initial instrument was only capable of magnifying objects by four hundred times, it demonstrated the principles of an electron microscope. Two years later, Ruska constructed an electron microscope that exceeded the resolution possible using an optical microscope.

Electron microscope constructed.

Reinhold Rudenberg, the research director of Siemens, had patented the electron microscope in 1931, although Siemens was doing no research on electron microscopes at that time. In 1937, Siemens began funding Ruska and Bodo von Borries to develop an electron microscope. Siemens also employed Ruska's brother Helmut to work on applications, particularly with biological specimens.

In the same decade Manfred von Ardenne pioneered the scanning electron microscope and his universal electron microscope.

Siemens produced the first commercial TEM in 1939, but the first practical electron microscope had been built at the University of Toronto in 1938, by Eli Franklin Burton and students Cecil Hall, James Hillier, and Albert Prebus.

Although modern electron microscopes can magnify objects up to two million times, they are still based upon Ruska's prototype. The electron microscope is an integral part of many laboratories. Researchers use them to examine biological materials (such as microorganisms and cells), a variety of large molecules, medical biopsy samples, metals and crystalline structures, and the characteristics of various surfaces. The electron microscope is also used extensively for inspection, quality assurance and failure analysis applications in industry, including, in particular, semiconductor device fabrication.

Types

Transmission Electron Microscope (TEM)

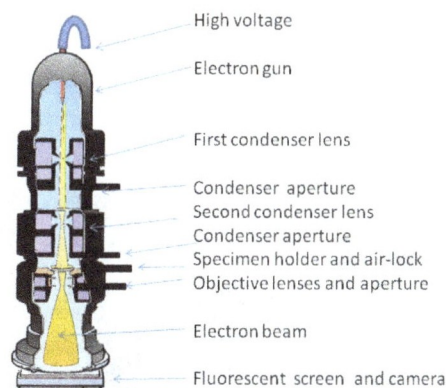

High voltage
Electron gun
First condenser lens
Condenser aperture
Second condenser lens
Condenser aperture
Specimen holder and air-lock
Objective lenses and aperture
Electron beam
Fluorescent screen and camera

Transmission Electron Microscope

Diagram of a transmission electron microscope.

The original form of electron microscopy, Transmission electron microscopy (TEM) involves a high voltage electron beam emitted by an electron gun, usually fitted with a tungsten filament cathode as the electron source. The electron beam is accelerated by an anode typically at +100keV (40 to 400 keV) with respect to the cathode, focused by electrostatic and electromagnetic lenses, and transmitted through a specimen that is in part transparent to electrons and in part scatters them out of the beam. When it emerges from the specimen, the electron beam carries information about the structure of the specimen that is magnified by the objective lens system of the microscope. The spatial variation in this information (the "image") is recorded by projecting the magnified electron image onto a fluorescent viewing screen coated with a phosphor or scintillator material such as zinc sulfide. The image can be photographically recorded by exposing a photographic film or plate directly to the electron beam, or a high-resolution phosphor may be coupled by means of a fiber optic light-guide to the sensor of a CCD (charge-coupled device) camera. The image detected by the CCD may be displayed on a monitor or computer.

Resolution of the TEM is limited primarily by spherical aberration, but a new generation of aberration correctors have been able to partially overcome spherical aberration to increase resolution. Software correction of spherical aberration for the High Resolution TEM HRTEM has allowed the production of images with sufficient resolution to show carbon atoms in diamond separated by only 0.89 ångström (89 picometers) and atoms in silicon at 0.78 ångström (78 picometers) at magnifications of 50 million times. The ability to determine the positions of atoms within materials has made the HRTEM an important tool for nano-technologies research and development.

Scanning Electron Microscope (SEM)

Unlike the TEM, where electrons of the high voltage beam form the image of the specimen, the Scanning Electron Microscope (SEM)produces images by detecting low energy secondary electrons which are emitted from the surface of the specimen due to excitation by the primary electron beam. In the SEM, the electron beam is rastered across the sample, with detectors building up an image by mapping the detected signals with beam position.

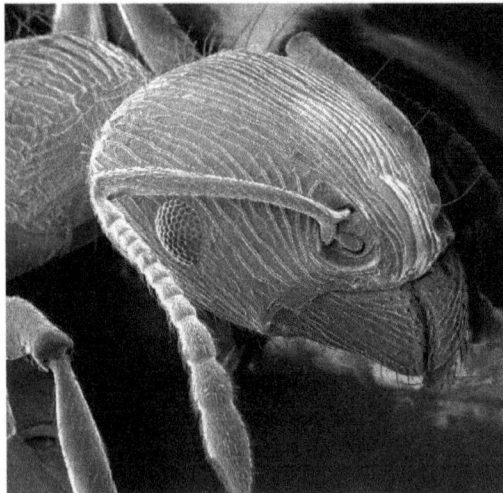

An image of an ant from a scanning electron microscope

Generally, the TEM resolution is about an order of magnitude greater than the SEM resolution, however, because the SEM image relies on surface processes rather than transmission it is able to image bulk samples and has a much greater depth of view, and so can produce images that are a good representation of the 3D structure of the sample.

Reflection Electron Microscope (REM)

In the Reflection Electron Microscope (REM) as in the TEM, an electron beam is incident on a surface, but instead of using the transmission (TEM) or secondary electrons (SEM), the reflected beam of elastically scattered electrons is detected. This technique is typically coupled with Reflection High Energy Electron Diffraction and Reflection high-energy loss spectrum (RHELS). Another variation is Spin-Polarized Low-Energy Electron Microscopy (SPLEEM), which is used for looking at the microstructure of magnetic domains.

Scanning Transmission Electron Microscope (STEM)

The STEM rasters a focused incident probe across a specimen that (as with the TEM) has been thinned to facilitate detection of electrons scattered through the specimen. The high resolution of the TEM is thus possible in STEM. The focusing action (and aberrations) occurs before the electrons hit the specimen in the STEM, but afterward in the TEM. The STEM's use of SEM-like beam rastering simplifies annular dark-field imaging, and other analytical techniques, but also means that image data is acquired in serial rather than in parallel fashion.

Sample Preparation

An insect coated in gold for viewing with a scanning electron microscope.

Materials to be viewed under an electron microscope may require processing to produce a suitable sample. The technique required varies depending on the specimen and the analysis required:

- Chemical Fixation for biological specimens aims to stabilize the specimen's mobile macromolecular structure by chemical crosslinking of proteins with aldehydes such as formaldehyde and glutaraldehyde, and lipids with osmium tetroxide.

- Cryofixation: freezing a specimen so rapidly, to liquid nitrogen or even liquid helium temperatures, that the water forms vitreous (non-crystalline) ice. This preserves the specimen in a snapshot of its solution state. An entire field called cryo-electron microscopy has branched from this technique. With the development of cryo-electron microscopy of vitreous sections (CEMOVIS), it is now possible to observe virtually any biological specimen close to its native state.

- Dehydration: freeze drying, or replacement of water with organic solvents such as ethanol or acetone, followed by critical point drying or infiltration with embedding resins.

- Embedding, biological specimens: infiltration of the tissue with a resin such as Araldite epoxy or acrylic resin followed by ultra-thin sectioning and staining.

- Embedding, materials: After embedding in resin, the specimen is usually ground and polished to a mirror-like finish using ultra-fine abrasives. The polishing process must be performed carefully to minimize scratches and other polishing artifacts that reduce image quality.

- Sectioning: produces thin slices of specimen, semitransparent to electrons. These can be cut on an ultramicrotome with a diamond knife to produce ultrathin slices about 90 nm thick. Glass knives are also used because they can be made in the lab and are much cheaper.

- Staining: uses heavy metals such as lead, uranium or tungsten to scatter imaging electrons and thus give contrast between different structures, since many (especially biological) materials are nearly "transparent" to electrons (weak phase objects). In biology, specimens are usually stained "en bloc" before embedding and also later stained directly after sectioning by brief exposure to aqueous (or alcoholic) solutions of the heavy metal stains.

- Freeze-fracture or freeze-etch: a preparation method particularly useful for examining lipid membranes and their incorporated proteins in "face on" view. The fresh tissue or cell suspension is frozen rapidly (cryofixed), then fractured by simply breaking or by using a microtome while maintained at liquid nitrogen temperature. The cold fractured surface (sometimes "etched" by increasing the temperature to about -100°C for several minutes to let some ice sublime) is then shadowed with evaporated platinum or gold at an average angle of 45° in a high vacuum evaporator. A second coat of carbon, evaporated perpendicular to the average surface plane is often performed to improve stability of the replica coating. The specimen is returned to room temperature and pressure, then the extremely fragile "pre-shadowed" metal replica of the fracture surface is released from the underlying biological material by careful chemical digestion with acids, hypochlorite solution or SDS detergent. The still-floating replica is thoroughly washed from residual chemicals, carefully fished up on EM grids, dried then viewed in the TEM.

- Ion Beam Milling: thins samples until they are transparent to electrons by firing ions (typically argon) at the surface from an angle and sputtering material from the surface. A subclass of this is Focused ion beam milling, where gallium ions are used to produce an electron transparent membrane in a specific region of the sample, for example through a device within a microprocessor. Ion beam milling may also be used for cross-section polishing prior to SEM analysis of materials that are difficult to prepare using mechanical polishing.

- Conductive Coating: An ultrathin coating of electrically-conducting material, deposited either by high vacuum evaporation or by low vacuum sputter coating of the sample. This is done to prevent the accumulation of static electric fields at the specimen due to the electron irradiation required during imaging. Such coatings include gold, gold/palladium, platinum, tungsten, graphite etc. and are especially important for the study of specimens with the scanning electron microscope. Another reason for coating, even when there is more than enough conductivity, is to improve contrast, a situation more common with the operation of a FESEM (field emission SEM). When an osmium coater is used, a layer far thinner than would be possible with any of the previously mentioned sputtered coatings is possible.

Disadvantages

Pseudocolored SEM image of the feeding basket of Antarctic krill. Real electron microscope images do not carry any color information, they are greyscale. The first degree filter setae carry in v-form two rows of second degree setae, pointing towards the inside of the feeding basket. The purple ball is one micrometer in diameter. To display the total area of this structure one would have to tile this image 7500 times.

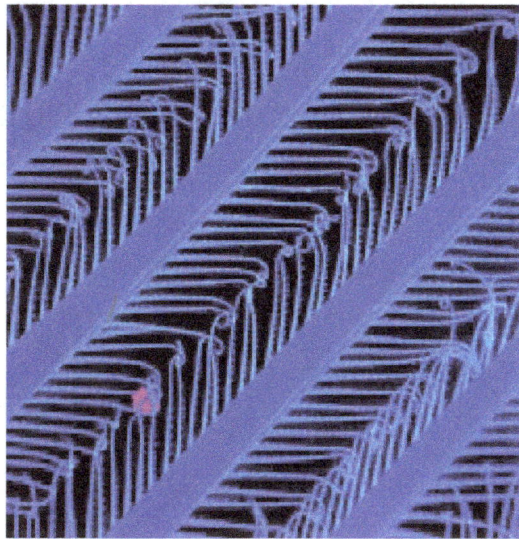

Electron microscopes are expensive to build and maintain, but the capital and running costs of confocal light microscope systems now overlaps with those of basic electron microscopes. They are dynamic rather than static in their operation, requiring extremely stable high-voltage supplies, extremely stable currents to each electromagnetic coil/lens, continuously-pumped high- or ultra-high-vacuum systems, and a cooling water supply circulation through the lenses and pumps. As they are very sensitive to vibration and external magnetic fields, microscopes designed to achieve high resolutions must be housed in stable buildings (sometimes underground) with special services such as magnetic field canceling systems. Some desktop low voltage electron microscopes have TEM capabilities at very low voltages (around 5 kV) without stringent voltage supply, lens coil current, cooling water or vibration isolation requirements and as such are much less expensive to buy and far easier to install and maintain, but do not have the same ultra-high (atomic scale) resolution capabilities as the larger instruments.

The samples largely have to be viewed in vacuum, as the molecules that make up air would scatter the electrons. One exception is the environmental scanning electron microscope, which allows hydrated samples to be viewed in a low-pressure (up to 20 Torr/2.7 kPa), wet environment.

Scanning electron microscopes usually image conductive or semi-conductive materials best. Non-conductive materials can be imaged by an environmental scanning electron microscope. A common preparation technique is to coat the sample with a several-nanometer layer of conductive material, such as gold, from a sputtering machine; however, this process has the potential to disturb delicate samples.

Small, stable specimens such as carbon nanotubes, diatom frustules, and small mineral crystals (asbestos fibers, for example) require no special treatment before being examined in the electron microscope. Samples of hydrated materials, including almost all biological specimens have to be prepared in various ways to stabilize them, reduce their thickness (ultrathin sectioning) and increase their electron optical contrast (staining). There is a risk that these processes may result in artifacts, but these can usually be identified by comparing the results obtained by using radically different specimen preparation methods. It is generally believed by scientists working in the field that as results from various preparation techniques have been compared and that there is no reason that they should all produce similar artifacts, it is reasonable to believe that electron microscopy features correspond with those of living cells. In addition, higher-resolution work has been directly compared to results from X-ray crystallography, providing independent confirmation of the validity of this technique. Since the 1980s, analysis of cryofixed, vitrified specimens has also become increasingly used by scientists, further confirming the validity of this technique.

Electron Microscopy Application Areas

Semiconductor and data storage	Research
• Circuit edit	• Electron beam induced deposition
• Defect analysis	• Materials qualification
• Failure analysis	• Materials and sample preparation
Biology and life sciences	• Nanoprototyping
	• Nanometrology
• Cryobiology	• Device testing and characterization
• Protein localization	**Industry**
• Electron tomography	
• Cellular tomography	• High-resolution imaging
• Cryo-electron microscopy	• 2D & 3D micro-characterization
• Toxicology	• Macro sample to nanometer metrology
• Biological production and viral load monitoring	• Particle detection and characterization
• Particle analysis	• Direct beam-writing fabrication
• Pharmaceutical QC	• Dynamic materials experiments
• 3D tissue imaging	• Sample preparation
• Virology	• Forensics
• Vitrification	• Mining (mineral liberation analysis)
	• Chemical/Petrochemical

Flow Cytometry

Flow cytometry is a technology that is used to analyse the physical and chemical characteristics of particles in a fluid as it passes through at least one laser. Cell components are fluorescently labelled and then excited by the laser to emit light at varying wavelengths.

The fluorescence can be measured to determine various properties of single particles, which are usually cells. Up to thousands of particles per second can be analysed as they pass through the liquid stream. Examples of the properties measured include the particle's relative granularity, size and fluorescence intensity as well as its internal complexity. An optical-to-electronic coupling system is used to record the way in which the particle emits fluorescence and scatters incident light from the laser.

Three main systems make up the flow cytometer instrument and these are the fluidics, the optics and the electronics. The purpose of the fluidics system is to transport the particles in a stream of fluid to the laser beam where they are interrogated. Any cell or particle that is 0.2 to 150 μms in size can be analyzed. If the cells are from solid tissue, they require disaggregation before they can be analyzed. Although cells from plants, bacteria, yeast or algae are usually measured, other particles such as chromosomes or nuclei can also be examined. Some particles such as marine algae are naturally fluorescent, but in general, fluorescent labels are required to tag components of the particle. The section of the fluid stream that contains the particles is referred to as the sample core.

The optics system is made up of lasers which illuminate the particles present in the stream as they pass through and scatter light from the laser. Any flourescent molecules that are on the particle emit fluorescence, which is detected by carefully positioned lenses. Generally, the light scattered from up to six or more fluorescences is determined for two different angles. Optical filters and beam splitters then direct the light signals to the relevant detectors, which emit electronic signals proportional to the signals that hit them. Data can then be collected on each particle or event and the characteristics of those events or particles are determined based on their fluorescent and light scattering properties.

The electronics system is used to change the light signals detected into electronic pulses that a computer can process. The data can then be studied to ascertain information about a large number of cells over a short period. Information on the heterogeneity and different subsets within cell populations can be identified and measured. Some instruments have a sorting feature in the electronics system that can be used to charge and deflect particles so that certain cell populations can be sorted for further analysis.

The data are usually presented in the form of single parameter histograms or as plots of correlated parameters, which are referred to as cytograms. Cytograms may display data in the from of a dot plot, a contour plot or a density plot.

Applications

Flow cytometry is used to perform several procedures including:

- Cell counting

- Cell sorting

- Detection of biomarkers

- Protein engineering

Flow cytometry has numerous applications in science, including those relevant to healthcare. The technology has been widely used in the diagnosis of health conditions, particularly diseases of the blood such as leukemia, although it is also commonly used in the various different fields of clinical practice as well as in basic research and clinical trials.

Some examples of the fields this technology is used in include molecular biology, immunology, pathology, marine science and plant biology. In medicine, flow cytometry is a vital laboratory process used in transplantation, oncology, hematology, genetics and prenatal diagnosis. In marine biology, the abundance and distribution of photosynthetic plankton can be analysed.

Flow cytometry can also be used in the field of protein engineering, to help identify cell surface protein variants.

Cell Sorting

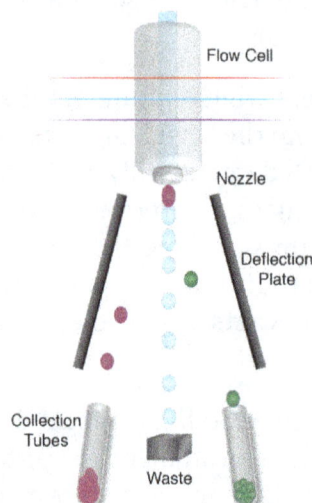

Cell sorting is the separation of cells from one another, based on physical or chemical properties. Cell-separation techniques are used to collect uniform populations of cells from tissues or fluids in which many different cell types are present. The collected cells can then be used for transplantation or scientific study. Common methods of separating cells include cloning, centrifugation, electrophoresis, magnetism, and antibody- or fluorescent-binding.

Cell sorting techniques fall into two general categories: bulk sorting and single cell sorting. In bulk cell sorting all of the target cells are collected in one sweep, whereas in single cell sorting every cell is individually analysed. There are multiple methods of bulk cell sorting: filtration, sedimentation, centrifugation, cell culture, and magnetic cell sorting. The main single cell sorting method is flow cytometry. While cell sorting can be very accurate, it is hard to say that a sorted cell population is "pure". Instead, the collected population is referred to as "enriched".

Magnetic Cell Sorting

Magnetic cell sorting is a bulk enrichment technique, and can be highly specific. Magnetic sorting using super paramagnetic nanoparticles introduce a high degree of specificity to a cell enrichment protocol. Super paramagnetic nanoparticles are made of a core of iron oxide, typically magnetite (Fe_3O_4), which is not innately magnetic, but becomes magnetized by an applied magnetic field. These particles or beads are coated with silica or a polymer surface to prevent clumping, and a well-chosen coating also provides a rich surface for the covalent attachment of functional groups and antibodies. The attachment of antibodies provides the super paramagnetic particles with specificity. The functionalized particles are incubated with the target cell solution, and the cells with surface antigens complementary to the antibodies will bind to form a cell-bead conjugate. The conjugates are enriched by magnetic cell separation. Magnetic cell sorting is a good choice when specificity is desired. Other bulk sorting methods such as filtration, sedimentation, and centrifugation cannot achieve such specificity. Magnetic sorting is rapid and efficient. Advanced bio magnetic separation systems can increase sort accuracy by providing standard curves and optical monitoring of the sorting process. Additionally, advanced separation systems are engineered to provide gentle and consistent magnetic forces throughout the working volume to increase cell viability.

Flow Cytometry fluorescence-based Cell Sorting

Cell sorting by flow cytometry analyzes each cell individually. This is an extremely powerful technique that can provide a large amount of information at once. Flow cytometry is an ideal quantification method for multiplex immunoassays. The cells are incubated with fluorophore-labeled antibodies before the sort. The antibodies are specific to surface antigens on target cells. Each antibody has a different emission wavelength and is uniquely identifiable. One method of precisely labeling antibodies with fluorophores is on-bead labeling with protein A.

After incubation with the labeled antibodies, the cell solution is sent through the flow cytometer. This machine guides the solution through a micron-sized nozzle one cell at a time. Each cell moves through a laser excitation area, where the laser excites the flurophores bound to the cell surface. The fluorescent emission is recorded, and the cell is directed either into a collection container or a discard container according to user-defined parameters. Multiple cell types can be enriched in a single run, and quantitative information about cell numbers and percent of total population are simultaneously recorded.

Flow cytometry is a highly informative method, and is the most precise cell sorting technique, but it is also very expensive. The machine itself is often prohibitively expensive, and requires a trained operator. Many institutions have a flow cytometry research core facility that charges by the hour for use of their machines, and these rates can also be quite high. For this reason many laboratories are turning back toward magnetic sorting techniques. Magnetic separation is specific, rapid, and efficient when care is taken to develop and finely tune a sorting strategy.

Plant Histological Techniques

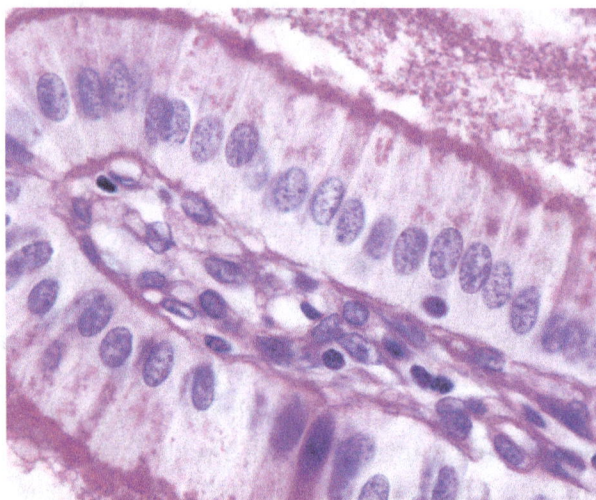

Plant histology is the branch of biology concerned with the composition and structure of plant tissues in relation to their specialized functions. Its aim is to determine how tissues are organized at all structural levels, from cells and intercellular substances to organs. It provides a realistic interpretation of morphology, physiology, and phylogeny of the structure of cells and tissues. A variety of techniques are used for histological studies by using various fixatives, stains, the use of microtome for preparing thin sections, light microscopy, electron microscopy, and X-ray diffraction.

To study tissues, the specimen is generally sliced into thin sections then contrast within tissues is induced using dyes, heavy metals, or fluorochromes. Specific staining is obtained by using a dye which has an affinity for a particular cell type or tissue element, or by the use of specific probes, such as labelled antibodies or labelled RNA or DNA probes. The basic requirements of histological studies are listed in Table below.

Table: Basic equipments and reagents required for histology

Basic Equipments	Basic Reagents
Glass vial	Ethanol
Rotator/shaker	Histoclear
Hotplate	Distilled water

Glass microscope slides	DPX permanent mounting medium
Cover slips	Aqueous mounting medium

Tissue Processing

The aim of tissue processing is to embed the tissue in a solid material firm enough to support the tissue and give it sufficient rigidity to enable thin sections to be cut, and yet soft enough not to damage the knife or tissue.

The five main stages in the preparation of histological slides are:

1. Fixation

2. Dehydration

3. Clearing

4. Sectioning

5. Staining

Fixation

The aim of fixation is to preserve the tissue in a state that most reflect the living cell. Choice of fixative is generally dependent on the tissue of interest as different fixatives better preserve particular tissue elements.

Aim of Fixation:

- To prevent autolysis and bacterial attack.

- To fix the tissues so they will not change their volume and shape during processing.

- To prepare tissue and leave it in a condition which allow clear staining of sections.

- To leave tissue as close as their living state as possible and no small molecules should be lost.

Conventional Chemical Fixation

Chemical fixatives are used to preserve tissue from degradation, and to maintain the structure of the cell and of sub-cellular components, such as cell organelles (e.g., nucleus, endoplasmic reticulum, mitochondria). There are many fixatives which have been developed over the years using mixture containing heavy metals or picric acid, but the most commonly used fixatives for general plant histology are buffered aldehyde and formalin/acid/alcohol mixtures (FAA).

Aldehyde fixatives usually contain a combination of paraformaldehyde and glutaraldehyde: paraformaldehyde rapidly penetrates the tissue while glutaraldehyde gives superior cross linking. Higher concentrations of glutaraldehyde improve morphological preservation but excessive cross linking destroys delicate antigenic sites on proteins. FAA fixes nucleic acids very well but gives poorer morphological preservation and makes the tissue hard and, therefore, more difficult to section.

Frozen Section Fixation

Frozen section is the rapid way to fix and mount histological sections. Fresh or fixed tissues are snap-frozen in liquid nitrogen, or with a high pressure jet of CO_2, and sectioned using a refrigeration device called a cryostat. The frozen tissue is sliced and the frozen slices are mounted on a glass slide and stained. The tissues are infiltrated in 40% sucrose prior to freezing and sectioning to help protect the tissue during freezing.

Advantages of fixation by frozen sections:

- Give better preservation of antigenicity

- Minimal exposure to fixative

- Not exposed to the organic solvents

- Disadvantages of fixation by frozen sections

- Lack morphological detail

- Possibility of biohazard

Dehydration

Dehydration removes fixative and water from the tissue and replaces them with dehydrating fluid. There are a variety of compounds used, many of which are alcohols. To minimize tissue distortion from diffusion currents, delicate specimens are dehydrated in a graded ethanol series from water through 10%-20%-50%-95%-100% ethanol.

In the paraffin wax method, following any necessary post fixation treatment, dehydration from aqueous fixatives is usually initiated in 60% -70% ethanol, progressing through 90%-95% ethanol, and then two or three changes of absolute ethanol before proceeding to the clearing stage. Examples of dehydrating agents used are ethanol, methanol, and acetone.

The duration of dehydration should be kept to the minimum consistent with the tissues being processed. Tissue blocks 1 mm thick should receive up to 30 minutes in each alcohol, blocks 5 mm thick require up to 90 minutes or longer in each change. Tissues may be held and stored indefinitely in 70% ethanol without harm.

Clearing

Clearing is the process of replacing the dehydrating fluid with another fluid that is totally miscible with both the dehydrating fluid and the embedding medium. The choice of a clearing agent depends upon the following:

- The type of tissues to be processed and the type of processing to be undertaken.

- Intended processing conditions, such as temperature, vacuum and pressure.

- Safety factors.

- Cost and convenience.

- Speedy removal of dehydrating agent.

- Ease of removal by molten paraffin wax.

- Minimal tissue damage.

Commonly used clearing agents include toluene, xylene, chloroform, benzene, etc.

Embedding and Sectioning

After tissues have been dehydrated and before they can be "sectioned" i.e. sliced very thinly, they must be secured in a very hard solid block in such a way that the hardened material secure all parts of the biological tissues and is transparent to the optical method used for viewing the finished samples. Generally, wax, polyethylene glycol (PEG), or resins (e.g. LR white) are used as embedding material for histology. During this process the tissue samples are placed into moulds along with liquid embedding material which is then hardened.

Paraffin wax is probably the most commonly used embedding material, being easy to section. It is a polycrystalline mixture containing solid hydrocarbons produced during the refining of coal and mineral oils. It is about two thirds of the density and slightly more elastic than dried protein. Wax and polyethylene glycol (PEG) are removable matrix; PEG is water soluble and requires lower infiltrating temperature than wax. Resins are not easily removable, however, they produce a much harder block and, therefore, thinner sections can be obtained.

Sectioning an embedded tissue sample is the next step necessary to produce sufficiently thin slices of sample that the detail of the microstructure of the cells/tissue can be clearly observed using microscopy techniques (either light microscopy or electron microscopy). Sectioning of embedded tissues depends on the type of microscopy that will be used to observe it and, hence, the thickness of sample required. In the case of samples to be studied using light microscopy, a steel knife mounted in a microtome may be used to cut 10µm thin sections, which are then mounted on a glass microscope slide. In case of samples to be studied using transmission electron microscopy, a diamond knife mounted in an ultra-microtome may be used to cut 50 nm thin sections, which are then mounted on a 3-millimeter-diameter copper grid. A microtome is a mechanical device utilized to slice biological specimens into very thin segments for microscopic examination. Most microtomes use a steel blade and are used to prepare sections of plant tissues for histology.

Staining

Most biological tissues have very little contrast, and cellular details are hard to discern with the ordinary light microscope. Staining is employed to give both contrast to the tissue as well as highlighting particular features of interest. Plant histologists have been staining tissues with natural dyes but today most dyes are synthetic. Dyes are extensively used in histology as they can enhance and improve the visibility of the specimen and often have an affinity for a specific tissue element, allowing quick and easy identification of specific cell types and cellular components. Although some dyes can be used on their own to stain tissues, specific dyes are often used with a counter stain to contrast various components within the tissues. Prior to staining, it is necessary to de-wax sections; resin sections can be stained directly.

Tolonium chloride (also known as toluidine blue) is a cationic dye that binds to negatively charged groups. An aqueous solution of this dye is blue, but different colors are generated when the dye binds with different anionic groups in the cell. For example, a pinkish purple color will appear when the dye reacts with carboxylated polysaccharides, such as pectic acid; green, greenish blue or bright blue with polyphenolic substances, such as lignin and tannins; and purplish or greenish blue with nucleic acids. Acridine orange is a nucleic acid selective fluorescent cationic dye that interacts with DNA and RNA by intercalation and electrostatic attractions, respectively. It is useful as a non-specific stain for backlighting conventionally stained cells. However, extreme care should be taken when using Acridine orange as it is carcinogenic. Haematoxylin is extracted from the heartwood of the logwood tree. It is one of the most commonly used stains in histology. It is generally utilized with a second counter stain, where haematoxylin as the primary stain for nuclei and contrasting or counterstaining with Orange G, safranin, or Fast Green.

Cell wall stains: Although cell walls can be visualized using general tissue stains such as Toluidine blue, there are a number of quick and simple methods which can be used to specifically stain individual cell wall components, such as pectin using Ruthenium red, cellulose using Calcofluor, lignin using Phloroglucinol or safranin and callose using Aniline blue, Resorcinol blue or Astra-blue.

Carbohydrates and starch stains: Total carbohydrates can be stained in tissue sections using the periodic acid Schiff's or PAS technique. The classic stain for starch uses iodine in potassium iodide.

Stains for cell viability: There are a number of fluorescent stains for assessing cell viability, however, the two fluorochromes, fluorescein diacetate (FDA) and propidium iodide are commonly used to detect living and dead cell, respectively, in tissues or cell suspensions. FDA is taken up by live cells and de-esterified to fluorescein, which fluoresces green with blue excitation. Conversely, propidium iodide is taken up by damaged or dead cells which fluoresces red with green excitation.

Stains for microorganisms in plant tissues: Bacteria are particularly difficult to detect with the light microscope as high magnification and high resolution optics are required. Fungi are generally detected in plant tissues using the dyes methyl blue or thionin. Bacteria can be detected using Gram staining technique. The Gram stain is widely performed on dried, heat fixed smears and gives blue stain with Gram positive microorganisms and red stain with Gram negative microorganisms.

Nucleic acid stains: The earliest method to stain nucleic acid (both DNA and RNA) in section is by using acridine orange, but this requires a fluorescence microscope. For bright field optics, methyl green-pyronin method is used. However, background pyronin staining is often a problem. DNA in nuclei, mitochondria and chloroplasts can be stained using DAPI (4',6-diamidino-2-phenylindole), Hoechst 33258, Hoechst 33342 or propidium iodide.

Lipid stains: Dyes like Sudan IV or Sudan black are generally used for staining of lipids. Nile blue is used preferentially to stain acid lipids, like phospholipids and works best on fresh tissues.

Nanotechnology in Plant

Nanotechnology is defined especially as growing and exciting technology at the scale of one-billionth of a meter sweeping away the barriers between the physics, chemistry and biology. Nanotechnology is the design, characterization, production and application of structures, devices and systems by controlling shape and size at nanometer scale. Nanotechnology in biomedical research has emerged as an interdisciplinary science that has quickly found its own niche in clinical methodologies including imaging, diagnostic, and therapeutics, drug delivery and tissue engineering. Nano medicine can design, build, manipulate, and optimize biological components at the Nano scale level. This includes the applications of Nano materials and the fabrication of Nano devices to be used in Nano diagnostic, Nano drug delivery and drug discovery.

Understanding the disease mechanisms of complex biological systems is still a significant challenge. Biological systems consist of hundreds of thousands of genes and proteins which are very hard to identify and whose behaviour is difficult to correlate, understand and predict. Synthetic biology, in combination to classical methods, is recently emerging as an alternative method. Individual mechanisms operating at various stages of the disease like initial, intermediate and advanced need further study to propose appropriate therapeutic intervention.

Nano particles (NPs) use their optical scattering properties for imaging and diagnostics, and their photo thermal properties for various types of therapies. The situation was improved by using active molecular targeting with cell-specific molecules (peptides, antibodies) attached to NPs and coupling to cognate receptors at the membranes of specific target (diseased) cells.

Agriculture and the Environment

Agriculture is the largest interface between humans and the environment, and is a major cause of climate change and ecosystem degradation. In particular, fertilizer use leads to fundamental changes in the pools Fertilizer utilization to supplement soil nutrients, to promote plant growth and to increase crop productivity and food quality is prevalent in modern agriculture. As a result, crop production and global food security are highly dependent on fertilizers input to agricultural lands. The selection and deployment of aims in stressed ecosystems therefore requires concerted research and technology development.

Pesticides use has dramatic consequences both in developed and developing countries. Sustainable agriculture aims at long term maintenance of natural resources and agricultural productivity with minimal adverse impact on the environment. Pesticide chemicals may induce oxidative stress leading to generation of free radicals and alterations in antioxidants or Oxygen Free Radical (OFR) scavenging enzymes. Synthetic or fumigant pesticides used for plant protection and pests controlling in stores usually bring about resistance in these pests.

Nanoparticles in Controlling the Plant Diseases

Today, application of agricultural fertilizers, pesticides, antibiotics, and nutrients is typically by spray or drench application to soil or plants, or through feed or injection systems to animals. Delivery of pesticides or medicines is either provided as "preventative" treatment, or is provided once the disease causing organism has multiplied and symptoms are evident in the plant.

In this context, nanotechnologies offer a great opportunity to develop new products against pests.

Nanotechnology improves their performance and acceptability by increasing effectiveness, safety, patient adherence, as well as ultimately reducing health care costs.

Nano scale devices are envisioned that would have the capability to detect and treat an infection, nutrient deficiency, or other health problem, long before symptoms were evident at the macro-scale. This type of treatment could be targeted to the area affected with a greater awareness of the hazards associated with the use of synthetic organic insecticides, there has been an urgent need to explore suitable alternative products for pest control.The broad application of Molecular Biology revolutionized the field of Diagnostics.

Today, Nanomaterials have been designed for a variety of biomedical and biotechnological applications, including biosensors, enzyme encapsulation. Nanotechnology is based on the introduction of novel Nano-materials which can result in revolutionary new structures and devices using extremely biological sophisticated tools to precisely position molecule. Nanoparticles technology has emerged as a strategy to tackle developing new materials and selecting appropriate materials for each specific treatment, other factors need to be optimally selected in order to design better targeted Nano particles. These factors include the particles size, shape, sedimentation, drug encapsulation efficacy, desired drug release profiles, distribution in the body, circulation, and cost.

Development of targeted drug delivery will improve therapeutic efficacy through reductions in drug dosing intervals, and diminished toxicities. The overall goal of this imaging Nanoparticles is to reduce the number of unnecessary problems in agriculture.

Nanoparticles mediated plant transformation has the potential for genetic modification of plants for further improvement. Specifically, application of Nanoparticles technology in plant pathology targets specific agricultural problems in plant–pathogen interactions and provide new ways for crop protection. Herein we reviewed the delivery of Nano particulate materials to plants and their ultimate effects which could provide some insights for the safe use of this novel technology for the improvement of crops.

Many of the preparation methods of Nanoparticles can be modified to create Nano structured films and Nano composites, although some types of nanostructures require completely novel approaches.

Carbon Nanotubes

Vertically-aligned multi-walled carbon Nanotubes (VACNTs) are arousing interest from researchers in biomedical area due to their exceptional combination of mechanical properties, chemical properties, and biocompatibility. Carbon Nanotubes (CNT) and functionalized fullerenes Bucky balls with bio-recognition properties provide tools at a scale, which offers a tremendous opportunity to study biochemical processes and to manipulate living cells at the single molecule level. Studies of this type can provide disease-gene-damage prone information for exploring DNA-safe therapeutics.

Carbon Nanotubes (CNTs) have become attractive electronic materials to date and their applications in future electric circuits and bio-sensing chips. CNT as vehicle to deliver desired molecules into the seeds during germination that can protect them from the diseases. Since it is growth promoting, it will not have any toxic or inhibiting or adverse effect on the plant.

Mesoporous Silica Nanoparticles

Nanoparticles can serve as 'magic bullets', containing herbicides, chemicals, or genes, which target particular plant parts to release their content. Mesoporous silica Nanoparticles (MSNs) has attracted the attention of several scientists over the last decade due to their potential applications. Among the main features of mesoporous materials is the high surface area, pore volume and the highly ordered pore network which is very homogeneous in size.

Mesoporous silica Nanoparticles (MSNs) have been extensive investigated as a drug delivery system. It is well know that MSNs possess excellent properties such high specific area, high pore volume, tunable pore structures and physicochemical stability. In the beginning MSNs were used for controlled delivery of various hydrophilic or hydrophobic active agents. Later advances in the MSNs surface properties such as surface functionalization and PEG ylation rendered them as a promising drug delivery.

Mesoporous silica Nanoparticles (MSN) helps in delivering DNA and chemicals into isolated plant cells. MSNs are chemically coated and serve as containers for the genes delivered into the plants. The coating triggers the plant to take the particles through the cell walls. It was found that MPS/DNA complexes showed enhanced transfection efficiency through receptor-mediated endocytosis via mannose receptors. These results indicate that MPS can be employed in the future as a potential gene carrier to antigen presenting cells.

Nanosensors

Although biosensors have been around since glucose monitors were commercialized in the 1970s, the transition of laboratory research and innumerable research papers on biosensors into the world of commerce has lagged.

Application of Nanoscale materials for electrochemical biosensors has been grown exponentially due to high sensitivity and fast response time. In these applications, effective immobilization of biomolecules without altering bioactivity is the key in construction of stable and wellstructured electrode materials for biosensor platform.

The developed biosensor system is an ideal tool for online monitoring of organophosphate pesticides and nerve agents. Bioanalytical Nanosensors are utilized to detect and quantify minute amounts of contaminants like viruses bacteria, toxins bio-hazardous substances etc. in agriculture and food systems. Most analysis of these toxins is still conducted using conventional methods; however, biosensor methods are currently being developed as screening tools for use in field analysis.

Nanoemulsion

The term 'Nanoemulsion' has been widely used to describe the complex systems consisting of oil phase, surfactant and water, which are optically isotropic and kinetically stable colloidal solution with droplet size in the range of 20-200 nm . Currently, Nanoemulsion are becoming the subject of many studies due to their wide range of particle sizes in Nanoscale, and this has contributed to more branches of potential uses and applications. Nanoemulsion was characterized for particle size viscosity, surface morphology and refractive index.

Nanoparticles suspensions very often present a physicochemical instability during their storage. In order to overcome this lack of stability and facilitate the handling of these colloidal systems, the water elimination from the aqueous dispersions to obtain a dry solid form appears as the most promising strategy.

Nano-emulsions, as non-equilibrium systems, present characteristics and properties which depend not only on composition but also on the preparation method. Nano-emulsions can encapsulate functional ingredients within their droplets, which can facilitate a reduction in chemical degradation.

Silver Nanoparticles

Silver Nanoparticles are appearing with ever-increasing frequency in consumer products, with over 300 self-identified nanosilver containing products on markets. These include dispersions and powders marketed as antimicrobials As novel nanosilver is incorporated into an increasing number of products subject to FDA regulation, questions about formulation, pyrogenicity, sterility, and sterilization procedures are emerging Since the size, shape and composition of silver Nanoparticles can have significant effect on their efficacy, extensive research has gone into synthesizing and characterizing silver Nanoparticles.

Silver Nanoparticles have also attracted much attention due to their diminutive size and novel material properties. With their nanometer scale size, which is responsible for different properties concerning the bulk material renders them suitable for applications. Therefore, many approaches have been used to prepare silver Nanoparticles for a rapidly growing list of catalysis, electronic, non-linear optics and biomaterial applications.

Nanosilver is used in agriculture to a wide extent because of its specific properties. A number of studies are conducted on the reaction of plants after their contact with nanosilver obtained by chemical reduction. Nano molecular silver solution reduced the incidence of root diseases. These examples demonstrate that the use of a colloidal nanosilver solution may considerably improve the growth and health of various plants.

Nanoparticles Mediated Nonviral Gene Delivery

Gene delivery systems are an important area in the field of genetic nanomedicine. Gene delivery involves the transport of genes, which requires a transport vehicle referred to as a vector. Possible vectors include viral "shells" or lipid spheres (Liposomes), which have properties that allow them to be incorporated into host cells.

Peptides and proteins have become the drugs of choice for the treatment of numerous diseases as a result of their incredible selectivity and their ability to provide effective and potent action. These studies suggest that research should be focused on designing a drug with an enhanced permeability and retention (EPR) effect. Nanoconjugate is being developed for non-invasive detection of gene expression in cells.

Polymer based gene transfer: Non-viral gene medicines have emerged as a potentially safe and effective gene therapy method for the treatment of a wide variety of acquired and genetic diseases.

An important advantage of polymer-based gene delivery systems over viral transfection systems is that transient gene expression without the safety concerns can be achieved. In addition to the polymeric systems to deliver DNA, therapeutic ultrasound is potentially useful because ultrasound energy can be transmitted through the body without damaging tissues and could be applied on a restricted area where the desired DNA is to be expressed.

Liposome gene transfer: The liposome-based gene transfer strategy is one of the most studied Nonviral gene delivery strategies. A liposomal delivery system requires a complete understanding of the physicochemical characteristics of the drug–liposome system. Many bacteria can control plant diseases by altering molecular processes leading to the production of pathogenicity and/or virulence factors by the pathogen.

Liposomes may offer several advantages as vectors for gene delivery into plant cells. Enhanced delivery of encapsulated DNA by membrane fusion, protection of nucleic acids from nuclease activity, targeting to specific cells, delivery into a variety of cell types besides protoplasts by entry through plasmodesmata. In Liposome based gene therapy there is no toxicity potential in humans and plants. Our results should stimulate efforts to develop plant-based technologies for the removal of pollutants from contaminated environments. Specific molecular changes have been suggested to be the reasons for the growth of gene therapics a liposomal delivery system requires a complete understanding of the physicochemical characteristics of the drug– liposome system.

Biobeads gene transfer: Micrometer-sized calcium alginate beads referred to as "bio-beads" that encapsulate plasmid DNA molecules carrying a reporter gene. In order to evaluate the efficiency of the bio-beads in mediating genetic transfection, protoplasts isolated from cultured tobacco cells. Transfection was up to 0.22% efficient. These results indicate that bio-beads have a possibility for efficient transformation in plants. Application of Nanoscale materials has been grown exponentially due to high sensitivity and fast response time. Hence focus will be on those systems whose response time must be within few milliseconds to a few seconds. Sometimes they may also cause some risk factors. Drug delivery systems with Liposomes and Nanoparticles have become very popular in nanotechnology sometimes these particles may also cause to microbial degradation.

A number of approaches are being developed to apply nanotechnology and particularly Nanoparticles to cleaning up soils contaminated with pesticides. To explore the benefits of applying nanotechnology to agriculture, the first stage is to work out the correct penetration and transport of the Nanoparticles into plants. This research is aimed to put forward a number of tools for the detection and analysis of core-shell magnetic Nanoparticles introduced into plants and to assess the use of such magnetic Nanoparticles in selected plant tissues.

References

- Nanotechnology-in-agriculture-2157-7439: omicsonline.org, Retrieved 28 June 2018
- Cell-sorting-techniques: sepmag.eu, Retrieved 09 April 2018
- What-is-Flow-Cytometry, life-sciences: news-medical.net, Retrieved 06 July 2018
- Electron-microscope-History: newworldencyclopedia.org, Retrieved 26 May 2018
- Microscopy, wmopen-biology-1: lumenlearning.com, Retrieved 12 March 2018

Chapter 7

Biofertilizers

A biofertilizer is a substance that consists of microorganisms, which is applied to plants to promote growth. The microbes act to colonize the rhizosphere and the plant's interior, and also increase the supply of primary nutrients to the plant. The aim of this chapter is to explore the use and production of biofertilizers, and its varied types such as phosphorus biofertilizers, compost biofertilizers and nitrogen biofertilizers.

Biofertilizers are defined as preparations containing living cells or latent cells of efficient strains of microorganisms that help crop plants' uptake of nutrients by their interactions in the rhizosphere when applied through seed or soil. They accelerate certain microbial processes in the soil which augment the extent of availability of nutrients in a form easily assimilated by plants.

Very often microorganisms are not as efficient in natural surroundings as one would expect them to be and therefore artificially multiplied cultures of efficient selected microorganisms play a vital role in accelerating the microbial processes in soil.

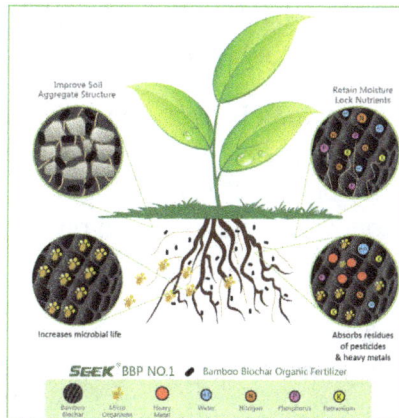

Use of biofertilizers is one of the important components of integrated nutrient management, as they are cost effective and renewable source of plant nutrients to supplement the chemical fertilizers for sustainable agriculture. Several microorganisms and their association with crop plants are being exploited in the production of biofertilizers. They can be grouped in different ways based on their nature and function.

Biofertilizers can be categorised in different ways based on their nature and function.

One simple broadly disseminated classification is as follows:

Nitrogen Biofertilizers

This group fixes nitrogen symbiotically. Nitrogen biofertilizers help to correct the nitrogen levels in the soil. Nitrogen is a limiting factor for plant growth because plants need a certain amount of nitrogen in the soil to thrive. Different biofertilizers have an optimum effect for different soils, so the choice of nitrogen biofertilizer to be used depends on the cultivated crop. Rhizobia are used for legume crops, Azotobacter or Azospirillum for non-legume crops, Acetobacter for sugarcane and blue-green algae and Azolla for lowland rice paddies.

Phosphorus Biofertilizers

Just like nitrogen, phosphorus is also a limiting factor for plant growth. Phosphorus biofertilizers help the soil to reach its optimum level of phosphorus and correct the phosphorus levels in the soil. Unlike nitrogen biofertilizers, the usage of phosphorus biofertilizers is not dependent on the crops cultivated on the soil. Phosphatika is used for all crops with Rhizobium, Azotobacter, Azospirillum and Acetobacter.

Compost Biofertilizers

Biofertilizers are also used for enrichment of your compost and for enhancement of the bacterial processes that break down the compost waste. Suitable biofertilizers for compost use are cellulolytic fungal cultures and Phosphotika and Azotobacter cultures. A 100% pure eco-friendly organic fertilizer is Vermi Compost: this organic fertilizer has nitrogen, phosphorus, potassium, organic carbon, sulphur, hormones, vitamins, enzymes and antibiotics, which helps to improve the quality and quantity of yield. It is observed that, due to continuous misuse of chemical fertilizers, the soil looses its fertility and becomes saline day by day. To overcome such problems, natural farming is the only remedy and Vermi compost is the best solution.

Another eco-friendly organic fertilizer which is prepared from sugar industry waste material that is decomposed and enriched with various plants and human-friendly bacteria and fungi is Biocompost. Biocompost consists of nitrogen, phosphate-solubilizing bacteria and various beneficial fungi like the decomposing fungus Trichoderma viridae, which protects plants from various soil-borne diseases and also helps to increase the soil fertility, resulting in a good quality product for farmers.

A more detailed classification of biofertilizers is as follows:

S.N	Groups		examples
Classification of Biofertilizers			
A	**N$_2$ fixing Biofertilizer**		
	1.	Free-living	Azotobacter, Clostridium, Anabaena, Nostoc,
		Symbiotic	Rhizobium, Anabaena azollae
	3.	Associative Symbiotic	Azospirillum
B	**P Solubilizing Biofertilizer**		
	1.	Bacteria	Bacillus subtilis, Pseudomonas striata
	2.	Fungi	Penicillium sp, Aspergillus awamori
C	**P Mobilizing Biofertilizers**		
	1.	Arbuscular Mycorrhiza	Glomus sp., Scutellospora sp..
	2.	Ectomycorrhiza	Laccaria sp., Pisolithus sp., Boletus sp., Amanita sp.
	3.	Ericoid Mycorrhiza	Pezizella ericae
D	**Biofertilizer for Micro nutrients**		
	1.	Silicate and Zinc solubilizers	Bacillus sp.
E	**Plant Growth Promoting Rhizobacteria**		
	1.	Pseudomonas	Pseudomonas fluorescence

Just to remind, biofertilizers are defined as biologically active products or microbial inoculants of bacteria, algae and fungi (separately or in combination), which may facilitate the biological nitrogen fixation for the benefit of plants. Biofertilizers also include organic fertilizers (manure, etc.), which are rendered in an available form due to the interaction of microorganisms or due to their association with plants.

Biofertilizers thus include the following:

- symbiotic nitrogen fixers, Rhizobium spp.;

- Non-symbiotic, free-living nitrogen fixers (Azotobacter, Azospirillum, etc.);

- algal biofertilizers (blue-green algae or blue-green algae in association with Azolla);

- phosphate-solubilising bacteria;

- mycorrhizae;

- Organic fertilizers.

The various biofertilizers are as follows:

- Nitrogen-fixing biofertilizers Nitrogen-fixing bacteria function under two types of conditions, symbiotically and as free-living (non-symbiotic) as well as associative symbiotic bacteria.

Free-Living Nitrogen-Fixing Bacteria

They live freely in the soil and perform nitrogen fixation. Some of them are saprotrophic, living on organic remains, e.g., Azotobacter, Bacillus polymyxa, Clostridium, Beijerinckia. They are further distinguished into aerobic and anaerobic forms.

The property of nitrogen fixation is also found in photoautotrophic bacteria, e.g., Rhizobium, Rhodopseudomonas, Rhodospirillum, Chromatium. Inoculation of soil with these bacteria helps in increasing the yield and cutting down on nitrogen fertilizers. For example, Azotobacter occurring in fields of cotton, maize, jowar and rice not only increases the yield, but also cuts down on nitrogen fertilizer to about 10–25 kg/ha. Its inoculant is available under the trade name of Azotobactrin.

Rhizobia are soil bacteria which are able to colonize the legume roots and fix the atmospheric nitrogen symbiotically. The morphology and physiology of rhizobia will vary from free-living conditions to the bactcroid of nodules. They are the most efficient biofertilizer as per the quantity of fixed nitrogen. There are seven genera that are highly specific in forming nodules in legumes, referred to as a cross-inoculation group.

Azotobacter is a genus of heterotrophic free-living nitrogen-fixing bacteria present in alkaline and neutral soils. It is aerobic in nature, recommended for non-leguminous crops like paddy, millets, cotton, tomato, cabbage and other monocotyledonous crops. Azotobacter also produces growth-promoting compounds. Azotobacter performs well if the soil organic matter content is high. Response to Azotobacter has been seen in rice, maize, cotton, sugarcane, pearl millet, vegetable and some plantation crops.

Free-Living Nitrogen-Fixing Cyanobacteria

A number of free-living cyanobacteria, or blue-green algae, have the property of nitrogen fixation, e.g., Anabaena, Nostoc, Aulosira, Totypothrix, Cylindrospermum, Stigonema. Cyanobacteria are photosynthetic microorganisms. Therefore, they add organic matter as well as extra nitrogen to the soil. These chlorophyll-containing prokaryotic organisms fix atmospheric nitrogen.

Aulosira fertilissima is considered to be the most active nitrogen fixer of rice fields. Cylindrospermum licheniformegrows in sugarcane and maize fields. Cyanobacteria are extremely low-cost biofertilisers. Phosphate, molybdenum and potassium are supplied additionally.

Loose Association of Nitrogen-Fixing Bacteria

This bacterial group live partly within the root and partly outside. There is a fair degree of symbiosis between the host and the bacteria. Hence, they are called associative symbiotic bacteria. Azospirillum is an important bacterium in this group, recommended for millets, grass, wheat, maize, sorghum, rice etc.

Symbiotic Nitrogen-Fixing Bacteria

They form a mutually beneficial association with the plants. The bacteria obtain food and shelter from plants. In return, they give to the plants part of their fixed nitrogen. The most important group of symbiotic nitrogen-fixing bacteria are rhizobia (Sg. rhizobium). They form nodules on the roots of legume plants. There are about a dozen Rhizobium species which form associations with the roots of different legumes, e.g. R. leguminosarum, R. lupini, R. trifolii, R. meliloti, R. phaseoli.

These bacteria, also called rhizobia, can live freely in the soil but cannot fix nitrogen except for a strain of cowpea Rhizobium. They develop the ability to fix nitrogen only when they are present inside the root nodules. In the nodule cells, bacteria (bacteroids) lie in groups surrounded by the membrane of the host cells, which is lined by a pink-red pigment called leghemoglobin. Presently cultures of Rhizobium specific for different crops are raised in the laboratory.

Frankia, a nitrogen-fixing mycelial bacterium (actinomycete), is associated symbiotically with the root nodules of several non-legume plants like Casuarina, Alnus (Alder) Myrica, Rubus etc. The leaves of a few plants (e.g., Ardisia) develop special internal cavities for providing space to symbiotic nitrogen-fixing bacteria, Xanthomonas and Mycobacterium. Such leaves are a constant source of nitrogen fertilizer to the soil.

Symbiotic Nitrogen-Fixing Cyanobacteria

Nitrogen-fixing cyanobacteria (blue-green algae) form symbiotic associations with several plants, e.g. cycad roots, liverworts, Azolla (fern), and lichenized fungi. Azolla is an aquatic floating fern, found in temperate climate suitable for paddy cultivation. The fern appears as a green mat over water, which becomes reddish due to excess anthocyanin pigmentation. The blue-green algae, cyanobacteria (Anabaena azollae), present as a symbiont with this fern in the lower cavities actually fixes atmospheric nitrogen.

Azolla pinnata is a small free-floating fresh water fern which multiplies rapidly, doubling every 5–7 days. The fern can coexist with rice plants because it does not interfere with their growth.

Anabaena azollae resides in the leaf cavities of the fern. It fixes nitrogen. A part of the fixed nitrogen is excreted in the cavities and becomes available to the fern. The decaying fern plants release this nitrogen for utilization by the rice plants. When a field is dried up at the time of harvesting, the fern functions as green manure, decomposing and enriching the field for the next crop.

Microphos Biofertilizers

They release phosphate from bound and insoluble states, e.g., Bacillus polymyxa, Pseudomonas striata, Aspergillusspecies.

Mycorrhiza

The mycorrhiza is a mutually beneficial or symbiotic association of a fungus with the root of a higher plant. The most common fungal partners of mycorrhiza are Glomus species. Mycorrhizal roots show a sparse or dense wooly growth of fungal hyphae on their surface. Root cap and root hairs are absent.

Mycorrhiza is a potential biofertilizer which mobilizes P, Fe, Zn, B and other trace elements. It supplies moisture from far-off inches and is ideal for long duration crops. It can be stored up to 2 years and is dry powder resistant.

Depending upon the residence of the fungus, mycorrhizae are of two types— ectomycorrhiza and endomycorrhiza.

Ectomycorrhiza (Ectotrophic Mycorrhiza)

The fungus forms a mantle on the surface of the root. Internally, it lies in the intercellular spaces of the cortex. The root cells secrete sugars and other food ingredients into the intercellular spaces that feed the fungal hyphae. The exposed fungal hyphae increase the surface of the root to several times. They perform several functions for the plant as follows:

- Absorption of water,

- Solubilisation of organic matter of the soil humus, release of inorganic nutrients, absorption and their transfer to root,

- Direct absorption of minerals from the soil over a large area and handing over the same to the root. Plants with ectomycorrhiza are known to absorb 2–3 times more of nitrogen, phosphorus, potassium and calcium,

- The fungus secretes antimicrobial substances which protect the young roots from attack of pathogens. Ectomycorrhiza occurs in trees such as Eucalyptus, oak (Quercus), peach, pine, etc. The fungus partner is generally specific. It belongs to Basidiomycetes.

Endomycorrhiza (Endotrophic Mycorrhiza)

Fewer fungal hyphae lie on the surface. The remaining live in the cortex of the root, mostly in the intercellular spaces with some hyphal tips passing inside the cortical cells, e.g., grasses, crop plants, orchids and some woody plants. At the seedling stage of orchids, the fungal hyphae also provide nourishment by forming nutrient-rich cells called pelotons. Intracellular growth occurs in order to obtain nourishment because, unlike ectomycorrhiza, the cortical cells do not secrete sugars in the intercellular spaces.

Vesicular Arbuscular Mycorrhizal (VAM) fungi possess special structures known as vesicles and arbusculars. VAM fungi are intercellular, obligate endosymbionts and, on establishment on the root system, act as an extended root system. Besides harvesting moisture from deeper and faraway nitches in the soil, they also harvest various micronutrients and provide them to the host plants. VAM facilitates the phosphorus nutrition by not only increasing its availability, but also increasing its mobility. VAM are obligate symbionts and improve the uptake of Zn, Co, P and H_2O. Its large-scale application is limited to perennial crops and transplanted crops. A single fungus may form a mycorrhizal association with a number of plants, e.g., Glomus.

The different types of biofertilizers are preparations made from natural beneficial microorganisms. They are safe for all plants, animals and human beings. Being beneficial to crops and natural nutrient cycles, they not only are environmentally friendly, but also help in saving of chemical inputs.

Main roles of biofertilizers:

- Make nutrients available.

- Make the root rhizosphere livelier.

- Growth-promoting substances are produced.

- More root proliferation.

- Better germination.

- Improve the quality and quantity of produce.

- Improve the fertilizer use efficiency.

- Higher biotic and abiotic stress tolerance.

- Improve soil health.

- Residual effect.

- Make the system more sustainable.

Liquid Biofertilizers

At present, biofertilizers are supplied to the farmers as carrier-based inoculants. As an alternative, liquid formulation technology has been developed which has more advantages than the carrier inoculants.

Benefits

The advantages of liquid biofertilizer over conventional carrier-based biofertilizers are listed below:

- Longer shelf-life – 12–24 months;

- No contamination;

- No loss of properties due to storage up to 45° C;

- Greater potential to fight with native population;

- Easy identification by typical fermented smell;

- Better survival on seeds and soil;

- Very easy to use by the farmer;

- High commercial revenues;

- High export potential.

Characteristics of Different Liquid Biofertilizers

Rhizobium

Physical features of Liquid Rhizobium Biofertilizer:

1. Dull white in colour;

2. No bad smell;

3. No foam formation, pH 6.8–7.5

Azospirillum

Physical features of liquid Azospirillum biofertilizer:

1. The colour of the liquid may be blue or dull white.

2. Bad odour confirms improper liquid formulation and may be considered as mere broth.

3. Production of yellow gummy colour materials confirms the quality product.

4. Acidic pH always confirms that there are no Azospirillum bacteria in the liquid.

Role of liquid Azospirillum under field conditions:

1. Stimulates growth and imparts green colour which is a characteristic of a healthy plant.

2. Aids utilization of potash, phosphorous and other nutrients.

3. Enhances the plumpness and succulence of fruits and increases the protein content.

Azotobacter

Physical features of liquid Azotobacter biofertilizer:

The pigment that is produced by Azotobacter in aged culture is melanin, which is due to oxidation of tyrosine by a copper-containing enzyme, tyrosinase. The colour can be seen in liquid forms. Some of the pigmentations are described below:

1. Produces brown-black pigmentation in liquid inoculum;

2. Produces yellow-light brown pigmentation in liquid inoculum;

3. Produces green fluorescent pigmentation in liquid inoculum;

4. Produces green fluorescent pigmentation in liquid inoculum;

5. Produces, pink pigmentation in liquid inoculum;

6. Produces less, gum-less, greyish-blue pigmentation in liquid inoculum;

7. Produces green-fluorescent pigmentation in liquid inoculum.

Acetobacter

These are sacharophillic bacteria associated with sugarcane, sweet potato and sweet sorghum plants. Acetobacterfixes 30 kg N/ha/year. This bacterium is mainly commercialized for sugarcane crops. It is known to increase the yield by 10–20 t/acre and sugar content by about 10–15 percent.

Advantages of the production technology of biofertilizers

Carrier-based	Liquid-based
Cheap	Longer shelf-life
Easier to produce	Easier to produce
Less investment	Temperature tolerant
	High cell counts
	Contamination-free
	More effective
	Product can be 100% sterile
Disadvantages	
Low shelf-life	High cost
Temperature sensitive	Higher investment for production unit
Contamination prone	
Low cell counts	
Less effective	
Automation difficult	

Methods of Application

Seed inoculation oR seed treatment

This is the most common practice of applying biofertilizers. In this method, the biofertilizers are mixed with 10% solution of jaggary. The slurry is then poured over the seeds spread on a cemented floor and mixed properly in a way that a thin layer is formed around the seeds. The treated seeds should be dried in the shade overnight and then they should be used. Generally, 750 grams of biofertilizer is required to treat the legume seeds for a one-hectare area.

Seedling Root Dip

The seedling roots of transplanted crops are treated for half an hour in a solution of biofertilizers before transplantation in the field. In this method, the seedlings required for one acre are inoculated using 2–2.5 kg biofertilizers. For this, a bucket having adequate quantity of water is taken and the biofertilizer is mixed properly. The roots of the seedlings are then dipped in this mixture so as to enable the roots to get inoculum. These seedlings are then transplanted. This method has been found very much suitable for crops like tomato, rice, onion, cole crops and flowers.

Main Field Application

This method is mostly used for fruit crops, sugarcane and other crops where localized application is needed. At the time of planting of fruit trees, 20 g of biofertilizer mixed with compost is to be added in the ring of one sapling. The same quantity of biofertilizer may be added in the ring soil of the seedling after it has attained maturity. Sometimes, biofertilizers are also introduced in the soil but this may require four to ten times more biofertilizers. Before use, the inoculants should be incubated with the desired amount of well decomposed granulated farmyard manure (FYM) for 24 hours. The FYM acts as nutrition medium and adjuvant (carrier) for biofertilizers.

Self-inoculation or Tuber Inoculation

This method is exclusively suitable for application of Azotobacter. In this method, 50 liters of water is taken in a drum and 4–5 kg of Azotobacter biofertilizer is added and mixed properly. Planting materials required for one acre of land are dipped in this mixture. Similarly, if we are treating potato, then the tubers are dipped in the mixture and planting is done after drying the materials in the shade.

Advantages and Disadvantages of Biofertilizer

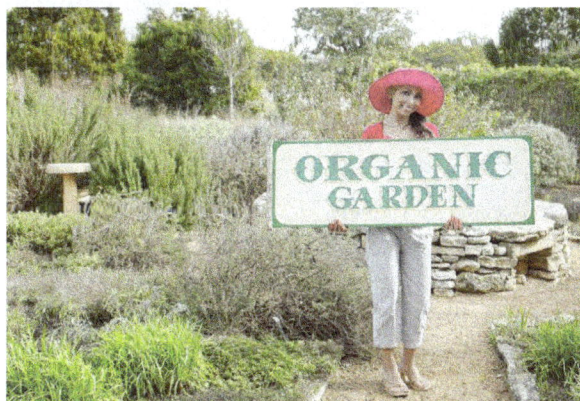

A biofertilizer is not just any organic fertilizer or manure. It consists of a carrier medium rich in live microorganisms. When applied to seed, soil or living plants, it increases soil nutrients or makes them biologically available. Biofertilizers contain different types of fungi, root bacteria or other microorganisms. They form a mutually beneficial or symbiotic relationship with host plants as they grow in the soil. Biofertilizers have many advantages and a few disadvantages.

Sustainability

Biofertilizers increase the nitrogen and phosphorus available to plants more naturally than other fertilizers.The different varieties available allow growers to tailor the microorganisms used to the needs of particular plants. Biofertilizers are simple to use, even for novice small growers. Biofertilizers do not pollute the soil or the environment, whereas chemical fertilizers often result in too much phosphate and nitrogen in the soil. The excess then leaches into lakes and streams through runoff. Waters decline in quality and suffer from overgrowth of algae and the death of fish.

Affordability

Biofertilizers reduce dependence upon expensive petroleum sources of chemical fertilizers. According to the "Journal of Phytology," demand for chemical fertilizers will exceed the supply by more than 7 million tons by 2020. The shortage of fossil fuels to produce chemical fertilizers may drive up prices beyond the reach of small users. Biofertilizers are a cheap, easy-to-use alternative to manufactured petrochemical products.

Improved Soil

Biofertilizers restore normal fertility to the soil and make it biologically alive. They boost the amount of organic matter and improve soil texture and structure. The enhanced soil holds water better than before. Biofertilizers add valuable nutrients to the soil, especially nitrogen, proteins and vitamins. They take nitrogen from the atmosphere and phosphates from the soil and turn them into forms that plants can use. Some species also produce natural pesticides.

Improved Plants

Biofertilizers increase yield by up to 30 percent because of the nitrogen and phosphorus they add to the soil. The improvement in soil texture and quality helps plants grow better during periods of drought. Biofertilizers help plants develop stronger root systems and grow better. Biofertilizers also reduce the effects of harmful organisms in the soil, such as fungi and nematodes. Plants resist stress better and live longer.

Disadvantages

Biofertilizers require special care for long-term storage because they are alive. They must be used before their expiration date. If other microorganisms contaminate the carrier medium or if growers use the wrong strain, they are not as effective. The soil must contain adequate nutrients for biofertilizer organisms to thrive and work. Biofertilizers complement other fertilizers, but they cannot totally replace them. Biofertilizers lose their effectiveness if the soil is too hot or dry. Excessively acidic or alkaline soils also hamper successful growth of the beneficial microorganisms; moreover, they are less effective if the soil contains an excess of their natural microbiological enemies. Shortages of particular strains of microorganisms or of the best growing medium reduce the availability of some biofertilizers.

Production of Biofertilizers

With day-by-day increasing the population, especially in developing countries, the stress on agriculture is also increasing continuously.

With the development, the land area under farming is not increasing but is further decreasing, this has posed extra burden on the agriculture. Therefore, the land available for agriculture should be economically utilized and maximum results is obtained.

Most of our agricultural lands are deprived of either one mineral or the other. These minerals are essential for the growth and development of plants. One of the nutrients for any type of plant is nitrogen. Nitrogen is a major element required by the plant for growth and development. The nitrogen is provided in the form of chemical fertilizer.

Such chemical fertilizers pose health hazards and pollution problem in soil besides these are quite expensive, bringing the cost of production much higher.

Therefore, bio-fertilizers are being recommended in place of chemical fertilizers. Bio-fertilizers are the formulations of living microorganisms which are able to fix atmospheric nitrogen in the available form for plants (nitrate form) either by living freely in the soil or associated symbiotically with plants.

Although nitrogen fixers are present in the soil, enrichment of soil with effective microbial strains is much beneficial for the crop yields. Use of composite bio-fertilizers can increase soil fertility.

It has been proved that bio-fertilizers are cost effective, cheap and renewable source to supplement the chemical fertilizers.

Production of Bio-Fertilizer

In order to meet the food requirements of ever increasing population, the nitrogen fertilizer requirement for crop production by 2000 A.D. was estimated to be about 11.4×106 tonnes. Biological nitrogen fixation can be the key to fill up this gap because of high cost and several other demerits of chemical fertilizers.

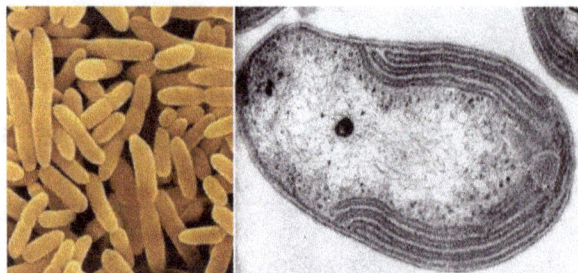

For production of a good and efficient bio-fertilizer, first of all an efficient nitrogen fixing strain is required, then its inoculum (the form in which the strain is to be applied in fields) is produced.

Packing, storing and maintenance are other aspects of bio-fertilizer production. While producing bio-fertilizers the standards laid down by BIS have also to be kept in mind for making the product authentic. Commercial production of bacteria, involved in the production of bio-fertilizer is shown in figure below.

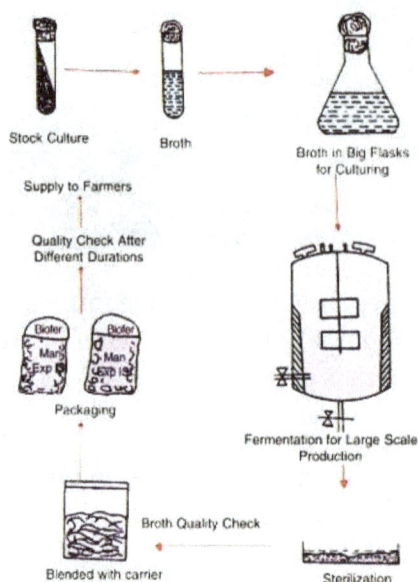

Criteria for Strain Selection

The efficient nitrogen fixing strain is evolved or selected in laboratory, maintained and multiplied on nutritionally rich artificial medium before inoculating the seed or soil. In soil, the strain has to survive and multiply to compete for infection site on roots against hostile environment in soil.

Steps for Preparing Bio-Fertilizer

The isolated strain is inoculated in small flasks containing suitable medium for inoculum production. The volume of the starter culture should be a minimum of 1% to obtain atleast 1×10^9 cells/ml. Now the culture obtained is added to the carrier for inoculant (bio-fertilizer) preparation.

Carriers carry the nitrogen fixing organisms to the fields. In some cases carrier is first sterilised and then inoculated, while in other cases it is first inoculated and then sterilised by UV irradiation.

The inoculum is now packed with 10^9-10^{10} viable cells per gram. Final moisture content should be around 40-60%. For large scale production of inoculum, culture fermenters are used.

Seed Pelleting

Direct seed coating with the gum arable or sugary syrup and useful nitrogen fixing strains especially the coating of rhizobia over specific host legume seeds are another method for obtaining fruitful results. As before, first of all the inoculum is prepared of the desired strain and then the seeds are inoculated by using either direct coating method or slurry method. Immediately after seed coating, $CaCO_3$ is added to sticky seeds.

The practice of seed inoculation dates back to 1896 when Voecher used this technique. In many soils the nodule bacteria are absent or are not adequate in either number or quality to meet the nitrogen requirements of the plants. Under these conditions, it is necessary to inoculate seeds or seedlings with highly effective rhizobia.

Inoculant Carriers

Most inoculants are the mixture of the broth culture and a finely milled, neutralized carrier material. Carrier is a substance having properties such as, non-toxicity, good moisture absorption capacity, free of lump forming material, easy to sterilize, inexpensive, easily available and good buffering capacity, so that it can prolong and maintain the growth of nitrogen fixing microorganisms which it is carrying.

The most frequently used carrier for inoculant production is peat.

A wide range of substitutes e.g. lignite, coal, charcoal, bagasse, filter mud, vermiculite, polyacrylamide, mineral soils, vegetable oils, etc. have been tested as alternative carriers.

Carrier processing e.g. mining, drying and milling are the most capital intensive aspects of inoculant (bio-fertiliser) production. First of all the carrier like peat is mined, drained and cleared off stones, roots, etc. Then, it is shredded and dried.

The peat is then passed through heavy mills. Material with a particle size of 10-40 µm is collected for seed coating. Peat with particle size of 500-1500 µm is used for soil implant inoculant. Carriers have to be neutralised by adding precipitated calcium carbonate (pH 6.5-7.0). After this, the carriers are sterilized for use as inoculants.

Quality Standards for Inoculants

Like every product, the bio-fertilizers should also follow certain standards. The inoculant should be carrier-based and should contain a minimum of 10^8 viable cells per gram of carrier on dry mass basis within 15 days of manufacture, and 10^7 within 15 days before the expiry date marked on the packet when the inoculant is stored at 25-30°C. The inoculant should have a maximum expiry period of 12 months from date of manufacture. The inoculant should not have any contaminant.

The contamination is one of the biggest problems faced by the bio-fertilizer industry. The pH of inoculant should be between 6.0 and 7.5. Each packet containing the bio-fertiliser should be marked with the information's e.g, name of product, leguminous crop for which intended, name and address of manufacture, type of carrier, batch or code number, date of manufacture, date of expiry, net quantity meant for net area and storage instructions. Each packet should also be marked with ISI (BIS) certification mark.

The inoculant (bio-fertilizer) should be stored in a cool place away from direct heat preferably at a temperature of 15°C. The bio inoculant should be packed in 50-75 μ low density polyethylene packets.

Two main methods of inoculation are currently being used (a) seed inoculation and (b) soil inoculation. The soil inoculation is done by delivering the inoculant directly into the sowing furrow with the seeds. Seed inoculation by pelleting or coating the seed with inoculant is the most popular methods.

Green Manuring

Green manuring is defined as a "farming practice where a leguminous plant which has derived enough benefits from its association with appropriate species of Rhizobium is ploughed into the field soil and then a non -legume is sown and allowed to get benefitted from the already present nitrogen fixer".

During the course of time, availability of chemical fertilizers decreased the significance of green manuring. In recent years, due to hike in price of chemical fertilizers, the practice of green manuring is reemphasized.

Some of the cultivated legumes and annual legumes such as Crotolaria juncea, C. striata, Cassia mimosoides, Cyamopsis pamas, Glycine wightii, Indigofera linifolia, Sesbania rostrata, Leucaena leucocephala, etc. contribute nitrogen.

In addition to nitrogen, green manuring provides organic matter, phosphorus, potassium besides minimising the pathogenic organisms in soil. The reclamation of "usar lands" can also be done by green manuring.

Algal and Other Bio-Fertilizers

Biological nitrogen fertilizers play a vital role to solve the problems of soil fertility, soil productivity and environmental quality. Anabaena azollae, a cyanobacterium lives in symbiotic association with the free floating water fern Azolla.

The symbiotic system Azolla-Anabaena complex is known to contribute 40-60 kg N ha-1 per rice crop. Anabaena azolle can grow photo autotrophically and fixes atmospheric nitrogen. The nitrogen fixing cyanobacteria such as A. azollae and variabilis when immobilized in polyurethane foam and sugar cane waste have significantly increased the nitrogen fixing activity and ammonia secretion.

The inoculation of cyanobacteria in nee crop significantly influenced the growth of rice crop by secretion of ammonia in flood water. The use of neem cake coupled with the inculation of Azolla greatly increased the nitrogen utilization efficiency in rice crop.

Besides Anabaena, other nitrogen fixing cyanobacteria like Aulosira, Calothrix, Hapalosiphon, Scytonema, Tolypothrix and Westiellopsis have been held responsible for the spontaneous fertility of the tropic rice fields.

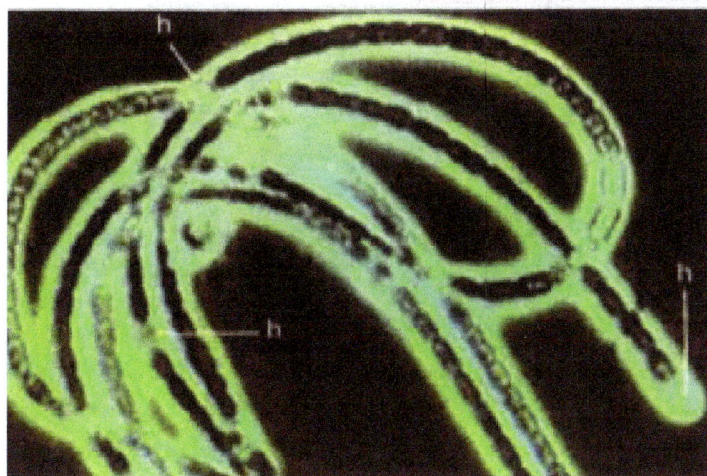

In addition to contributing N, the cyanobacteria add organic matter, secrete growth promoting substance like auxins and vitamins, mobilise insoluble phosphate and improve physical and chemical nature of the soil. Algalization has been shown to ameliorate the saline- alkali soils, help in the formation of soil aggregates, reduce soil compaction, and narrow C: N ratio.

These organisms enable the crop to utilize more of the applied nutrients leading to increased fertilizer utilising efficiency of crop plant. Most of the cyanobacteria act as supplements to fertilizer N contributing up to 30 kg N ha-1 season-1. The increase in the crop yield varies between 5-25 percent.

Mass Production of Cyanobacterial Biofertilizers

For outdoor cultivation of cyanobacterial biofertilizers, the regional specific strain should be used. In such practices, a mixture of 5 or 6 regionally acclimatized strains of cyanobacteria e.g. species of Anabaena, Aulosira, Cylindrospermum, Gloeotrichia, Nostoc, Plectonema, Tolypothrix etc. are generally used as starter inoculum.

The following Methods are used for Mass Cultivation:

- Cemented tank method,

- Shallow metal troughs method,

- Polythene lined pit method, and

- Field method.

The polythene lined method is most suitable for small and marginal farmers for the preparation of bio-fertilizer. In this method, small pits are prepared in field and lined with thick polythene sheets.

The mass cultivation of cyanobacteria is done by using any of the above four methods; the steps are given below

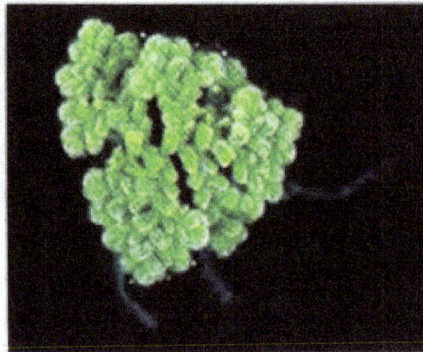

Azolla-Anabaena Symbiosis : Azolla, a water fern.

- Prepare the cemented tank, shallow trays of iron sheets or polythene lined pits in an open area. Width of tanks or pits should not be more than 1.5 m. This will facilitate the proper handling of culture.

- Transfer 2-3 kg soil and add 100 g superphosphate. Water the pit to about 10 cm height. Mix lime to adjust the pH. Add 2 ml of insecticide to protect the culture from mosquitoes. Mix well and allow to settle down soil particles.

- When water becomes clear, sprinkle 100 g starter culture on the surface of water.

- When temperature remains around 35-40°C during summer, optimum growth of cyano-bacteria is achieved. The water level is always maintained about 10 cm during the period.

- After drying, the algal mass (mat) is separated from the soil that forms flakes. During summer about 1 kg pure algal mat per m2 area is produced. It is collected, powdered, and packed in polythene bag and supplied to the farmers after sealing the packets.

- The algal flakes can be used as starter inoculum again.

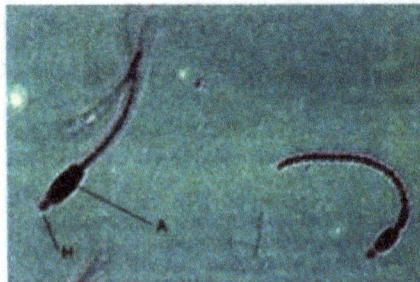

Terminal heterocysts (H) and subterminal akinetes (A) of Cylindrospermum, a cyanobacterium involved in mass production of biofertilizers.

Mass Cultivation of Azolla

The aquatic heterosporus fern contains endophytic cyanobacterium, Anabaena azollae in its leaf cavity. There are number of species of Azolla, namely A. caroliniana, A, filiculoides, A. maxicana,

A. nilotica, A. pinnata and A. rubra which are used as biofertilizer especially for paddy. For mass cultivation of Azolla, microplots (20 m2) are prepared in nurseries in which sufficient water (5-10 cm) is added.

For profuse growth of Azolla, 4-20 kg P2O5/ha is also amended. Optimum pH (8.0) and temperature (14-30°C) should be maintained. Finally, microplots are inoculated with fresh Azolla (0.5 to 0.4 kg/m2). An insecticide (Furadon) is used to check the insect's attack. After 3 weeks, the mat of Azolla is ready for harvest and the same microplot is inoculated with fresh Azolla to repeat the cultivation.

Azolla mat is harvested and dried to use as green manure.

There are two methods for its application in field

- Incorporation of Azolla in soil prior to rice cultivation, and

- Transplantation of rice followed by water draining and incorporation of Azolla.

However, reports from the IRRI, Manila (Philippines) revealed that growing of Azolla in rice field before rice transplantation increased the yield equivalent to that obtained from 30 kg/ha nitrogen as urea or ammonium phosphate.

Endophytic Nitrogen Fixers

Recently, several non-leguminous and particularly graminaceous species such as rice, wheat and forage grasses have registered tremendous interest in nitrogen fixation. Isolation of a number of diazotrophic bacteria such as Azospirillum, Herbaspirillum and Acetobacter is reported.

The term endophyte refers to microorganisms (bacteria and fungi) that colonize root interior of plants and live most of their life inside the plant tissue. Splitting the term endophyte into facultative and obligate was suggested to distinguish, respectively, strains that are able to colonize both the surface and root interior and to survive well in soil from those that do not survive well in soil but colonize the root interior and aerial parts.

Facultative Endophytic Diazotrophs

This group is composed of Azospirillum spp. and considered important with non-legume plants. Although A. lipoferum was the first species of the genus isolated by Tarrand. A. brasilense among all the seven known species is the best characterized at physiological and molecular levels.

Obligate Endophytic Diazotrophs

This group includes Acetobacter diazotrophicus (syn. Gluconacetobacter diazotrophicus) a nitrogen fixing bacterium clustered in the alpha sub-class of the proteobacteria, Azoarcus spp., Herbaspirillum spp. and a partially identified Burkholderia sp. are clustered in the beta sub-class of the proteobacteria.

Other Bacteria

Alcaligens, a diazotrophic member of this genus has been consistently isolated from the rhizosphere of wet rice land. Burkholderia, the other bacterium appears to have potential as rice inoculant. In the case of Klebsiella, substituted nitrogen fixation has been observed in rice inoculated with K. oxytoca or any other Klebsiella spp. that are considered as endophytes.

The diazotrophic nature of some members of the genus Pseudomonas is still a matter of debate. Nevertheless, several bacteria within it are clearly diazotrophic such as Pseudomonas diazotrophicus, P. flurorescens, P. saccharophila and P. stutzeri. Recently, several researchers have attempted to construct an artificial association between rhizobia and rice particularly with Azorhizobium caulinodans.

Isolation and Identification of Endophytes

For isolation and identification of natural diazotrophs from plant samples, root or stem, washed with sterile water, surface sterilized with 70% ethanol for 5 minutes and with sodium hypochlorite (2-5%) for 30 second, washed several times using sterile water. Sterilization of root and stem will be verified by rolling them on BMS agar plates.

Then homogenize the sample in a mortar and pestle in sterile phosphate buffer, saline 1% sugar solution and serially diluted and 0.1 ml sample transfer into vials containing 5-8 ml of respective semisolid media for the targeted bacterium with respective C sources with an initial pH of 6.0.

The number of diazotrophic populations is determined by the most probable number methods using a McCrady table. Vials with veil pellicles reaching the surface after incubation at 30°C with or without gas production and with positive reaction for acetylene reduction activity, show the presence of good endophytes.

Applications in Agriculture

Obligate endophytes have an enormous potential for use because of their ability to colonize the entire plant interior and establish themselves niches protected from oxygen or other inhibitory factors; thus their potential to fix nitrogen can be expressed maximally.

Recent studies in Brazil showed that the sugarcane varieties fix up to 80% nitrogen. It has been reported that wetland rice receives some nitrogen by endodiazotrophs. Tropical pasture grasses such as Brachiaria, Digitaria, Panicum and Paspalum spp. fix nitrogen.

Bio-Fertilizers Aiding Phosphorus Nutrition

Tropical soils are deficient in phosphorus. Further most of the microorganisms solubilize P and thus make it available for plant growth. It is estimated that in most tropical soils, 75% super phosphate applied is fixed and only 25% is available for plant growth.

There are some fungi such as Aspergillus awamori, Penicillium digitatum, etc. and bacteria like Bacillus polymyxa, Pseudomonas striata, etc. that solubilize unavailable form of P to available form. India has 250 mt of rock phosphate deposits. The cheaper source of rock phosphate like Mussoorie rock phosphate and Udaipur rock phosphate available in our country can be used along with phosphate solubilising microorganisms.

Vesicular-arbuscular mycorrhizal (VAM) fungi colonize roots of several crop plants. They are zygo-mycetous fungi belonging to the genera Glomus, Gigaspora, Acaulospora, Sclercystis, etc.

These are obligate symbionts and cannot be cultured on synthetic media. They help plant growth through improved phosphorus nutrition and protect the roots against pathogens. Nearly 25-30% of phosphate fertilizer can be saved through inoculation with efficient VAM fungi as reported by Bagyaraj.

Penicillium solubilizes unavailable form of P to available form.

Production of Mycorrhizal Bio-Fertilizer

Methods of inoculum production of mycorrhizal fungi differ with respects to their nature, depending upon types i.e., ectomycorrhizal or endomycorrhizal.

Ectomycorrhizal Fungi

In this case, the basidiospores, chopped sporocarps, sclerotia, pure mycelial culture, fragmented mycorrhizal roots or soil from mycorhizosphere region can be used as inoculum. The inoculum is mixed with nursery soil and seeds are sown thereafter.

Institute for mycorrhizal Research and Development, USA and Abbot Laboratories, USA have developed a mycelial inoculum of Pisolithus tinctorius in a mycelial vermiculite-peat moss substrate with trade name 'MycoRhiz' which is commercially available on large quantities.

VA Mycorrhizal Fungi

VA mycorrhiza can be produced on a large scale by pot culture technique. This requires the host plant mycorrhizal fungi and natural soil. The host plants which support large scale production of inoculum are sudan grass, strawberry, sorghum, maize, onion, citrus, etc.

The starter inoculum of VAM can be isolated from soil by wet sieving and decantation technique. VAM spores are surface sterilised and brought to the pot culture. Commonly used pot substrates are sand: soil (1:1, w/w) with a little amount of moisture.

There are two Methods of using the Inoculum

- Using a dried spore-root-soil to plants by placing the inoculum several centimetres below the seeds or seedlings,

- Using a mixture of soil- roots, and spores in soil pellets and spores are adhered to seed surface with adhesive,

Commercially available pot culture of VA mycorrhizal hosts grown under aseptic conditions can provide effective inoculum. Various types of VAM inocula are currently produced by Native Plants, Inc (NPI), Salt Lake City.

Nitrogen Biofertilizers

Nitrogen is an essential plant nutrient. Unfortunately, it is usually not present in soil at concentrations sufficient for agricultural production of commercial crops. It therefore must be provided to the crops in the form of fertilizer. Of the commercially important crops, cereals such as corn, wheat, rice and barley require particularly large amounts of fertilizer. For example, Azotobacter is used for the non-legume crops; Rhizobium is needed for the legume crops. Similarly blue green algae are needed to grow rice while Acetobacter is used to grow sugarcane. It means almost all the crops need different types of biofertilizers depending on their needs.

The use of nitrogen-fixing bacteria called biofertilization crops or inoculation and is used with some success worldwide, but the main problem is the low efficiency of inoculants, so that their use in the field has not been extensive. The production of a biofertilizer industry represents a much lower cost to produce a fertilizer. For example, one hectare of beans is about 500 pesos needed to fertilize and to inoculate the same surface with bacteria, without reducing the performance, about 10 pesos. Then, an affordable alternative and more environmentally appropriate for achieving the objective in the field is the improvement of atmospheric nitrogen fixation occurs in the cultivation of plants.

Azospirillum

Naturally occurring Azospirillum bacteria are free- living, aerobic, gram-negative, motile and nitrogen fixing. They generally prefer organic acids as carbon sources, e.g. malic or lactic acid, and fix nitrogen in the absence of a combined nitrogen source, under micro- aerophilic conditions (low oxygen tension). They also have relatively short shelf lives with low pectinolytic activity (the ability to break down pectin present in the cell walls of cereal plants so as to render the cell walls more permeable to minerals, hormones) and generally are not resistant to commonly used, commercial fungicides, herbicides and other pesticides or to antibiotics released by different soil microorganisms.

To overcome certain disadvantages associated with naturally-occurring Azospirillum strains, the novel strains having relatively improved survivability, or resistance to commonly used pesticides or increased pectinolytic activity or a combination of these traits. Such strains are able to enhance crop yields or reduce nitrogen fertilizer requirements, or both.

Since pectinolytic activity can be correlated with usefulness as biofertilizer, novel strains of Azospirillum having pectinolytic activity relative to the pectinolytic activity of naturally-occurring strains in the range from about 1.5:1 to 20:1 have been created. Presently, the preferred strains are of the species brasilense. Optionally, they are available as biologically pure, stable cultures.

The fertilizer composition may comprise at least 5% by total bacterial content of the Azospirillum brasilense; at least 30% by total bacterial content of the Azospirillum brasilense; or at least 50% by total bacterial content of the Azospirillum brasilense. The fertilizer composition may have the bacterial consortium comprise Azospirillum brasilense in combination with at least two other bacteria selected from the group consisting of Ochrobacterium tritici sp.; Ensifer adhaerens; Sinorhizobium sp; Enterobacter sp.; Zooglea sp.; Brevibacillus sp.; Bacillus cereus; and Agrobacterium tumefaciens; or at least two bacteria selected from the group consisting of Ochrobacterium tritici sp.; Ensifer adhaerens; Sinorhizobium sp; Enterobacter sp.; Zooglea sp.; Brevibacillus sp.; and Bacillus cereus.

Azotobacter

Azotobacter naturally fixes atmospheric nitrogen in the rhizosphere. There are different strains of Azotobacter each has varied chemical, biological and other characters. However, some strains have higher nitrogen fixing ability than others. Azotobacter uses carbon for its metabolism from simple or compound substances of carbonaceous in nature. Besides carbon, Azotobacter also requires calcium for nitrogen fixation. Similarly, a medium used for growth of Azotobacter is required to have presence of organic nitrogen, micro-nutrients and salt in order to enhance the nitrogen fixing ability of Azotobacter.

The bacterial fertilizer (biofertilizer) comprised of a suspension which contains cells of new strains of the micro-organisms Azotobacter vinelandii CECT 4534 and Azospirillum brasilense. Said bacterial fertilizer is a suspension of micro-organisms in a water medium. The micro-organisms are obtained by a process which comprises: a) plate culture of the original strain; b) treatment of the former colonies of the prior culture with another culture means; and c) culture of the isolated colonies of the prior step.

The advantages of the use of strains Azospirillum Azotobacter over the known strains are

1. The strains collected from Azospirillum Azotobacter and reproduce well at a greater range of pH and better than the known strains, supporting very alkaline media (pH = 10), thus avoiding problems of sterilization and contamination.

2. These strains are able to fix more N2 more easily and eliminate the ion NH4+ to the culture medium, while polysaccharides excreting more concentrated than known strains. These polysaccharides serve Azotobacter cells and Azospirillum as an energy reserve for further development on the ground while allowing them to adhere more easily to the roots of plants.

3. These variants Azospirillum Azotobacter and also have a greater capacity to assimilate the compounds formed by the exudates from the roots of plants, such as p-hydroxybenzoic acid, which is hardly used by other races.

4. The rate of reproduction and assimilation of the substrates is also higher in Azotobacter variants and strains Azospirillum known that, since the maximum dry weight attained by them are up to three times higher than the 52 hours of culture dipped to about 28-30 ° C.

5. The performance and effectiveness of these variants Azotobacter is far superior to known strains and the initial strain DSM 382, also synthesized with high gelling poly-saccharides (alginates).

Rhizobium

Producing a biofertilizer for plants with improved efficiency in nitrogen fixation has been the target of several research groups for several years because it represents economic benefits, agricultural and environmental important. Improving the process of nitrogen fixation could be obtained through the use of organisms such as Rhizobium bacteria genetically modified. However, despite the various strategies that have been conducted, researchers have reported limited success in its efforts to improve nitrogen symbiotic fixation.

The expression of construction in the Rhizobium strains tested were characterized by increased levels of expression of the enzyme complex nitrogenase in Rhizobium bacteria to provide a high transcription promoter region coupled to the full operon nifHDK (SEQ ID NO: 2, 3 and 4). It is important to mention that there is no report or strain, except for those covered by this application; they are able to increase the binding capacity of Rhizobium.

The potential use as biofertilizer Rhizobium agencies to contain the gene construct with which nitrogenase is overexpressed (pr. nifHDK c) is supported in field trials. Although the tests were performed on the bean crop, this is just one example of how to carry out the implementation of the invention. Additionally it can be applied in other legumes such as peanuts, soybeans, alfalfa, clover, lentils, beans, peas, etc.. and even in other crops such as sugar beets, wheat, maize and sorghum. In the construction of the strains always considered their possible use in the Mexican countryside and therefore meet all the requirements for approval of release of genetically modified organisms which marks the National Biosafety Committee, the competent authority in the matter.

Therefore, the main feature of Rhizobium strains containing this gene construct in association with a legume such as beans are reaching increased levels of seed production and nutrient content of the same and that in fixing more nitrogen, it is incorporated protein and amino acids in the seed in larger quantities.

Phosphorus Biofertilizers

Phosphorus biofertilizers are used to determine the phosphorus level in the soil. The need of phosphorus for the plant growth is also limited. Phosphorus biofertilizers make the soil get the required amount of phosphorus. It is not necessary that a particular phosphorus biofertilizers is used for a particular type of crop. They can be used for any types of the crops for example; Acetobacter, Rhizobium and other biofertilizers can use phosphotika for any crop type.

Phosphate Solubilizing Bacteria

Phosphate solubilizing bacteria (PSB) are beneficial bacteria capable of solubilizing inorganic phosphorus from insoluble compounds. P-solubilization ability of rhizosphere microorganisms is considered to be one of the most important traits associated with plant phosphate nutrition. It is generally accepted that the mechanism of mineral phosphate solubilization by PSB strains is associated with the release of low molecular weight organic acids, through which their hydroxyl and carboxyl groups chelate the cations bound to phosphate, thereby converting it into soluble forms. PSB have been introduced to the Agricultural community as phosphate Biofertilizer. Phosphorus (P) is one of the major essential macronutrients for plants and is applied to soil in the form of phosphate fertilizers. However, a large portion of soluble inorganic phosphate which is applied to the soil as chemical fertilizer is immobilized rapidly and becomes unavailable to plants. Currently, the main purpose in managing soil phosphorus is to optimize crop production and minimize P loss from soils. PSB have attracted the attention of agriculturists as soil inoculums to improve the plant growth and yield. When PSB is used with rock phosphate, it can save about 50% of the crop requirement of phosphatic fertilizer. \The use of PSB as inoculants increases P uptake by plants. Simple inoculation of seeds with PSB gives crop yield responses equivalent to 30 kg P2O5 /ha or 50 percent of the need for phosphatic fertilizers. Alternatively, PSB can be applied through fertigation or in hydroponic operations. Many different strains of these bacteria have been identified as PSB, including Pantoea agglomerans (P5), Microbacterium laevaniformans (P7) and Pseudomonas putida (P13) strains are highly efficient insoluble phosphate solubilizers. Recently, researchers at Colorado State University demonstrated that a consortia of four bacteria (sold commercially as Mammoth P), synergistically solubilize phosphorus at a much faster rate than any single strain alone.

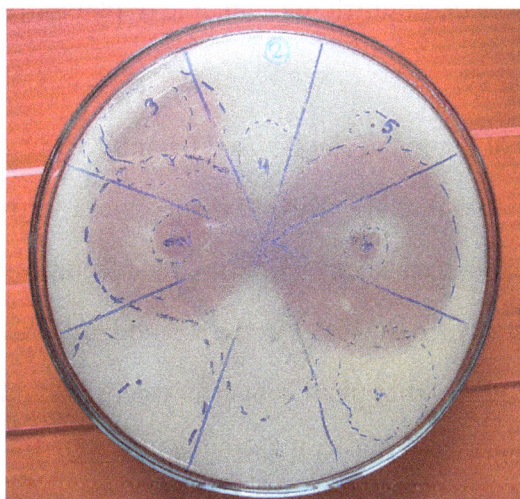

Phosphate solubilizing bacteria cultured in petri dish. The zone of n clearance can be clearly seen.

Additionally, phosphate (P) compounds are capable of immobilizing heavy metals, especially Pb, in contaminated environments through phosphate-heavy metal precipitation. However, most P compounds are not readily soluble in soils so it is not readily used for metal immobilization. Phosphate solubilizing bacteria (PSB) have the potential to enhance phosphate-induced immobilization of metals to remediate contaminated soil. However, there is a limit on the amount of phosphate which can be added to the environment due to the issue of eutrophication.

Phosphate is often adsorbed onto the surface of different type of minerals, for example iron containing minerals. Recent data suggest that bacteria growin under phosphorus starvation release iron-chelating molecules. Considering the geochemical interaction between these two elements, the authors suggest that some bacteria can dissolve iron-containing minerals in order to access the adsorbed phosphate.

Compost Biofertilizers

One of the best natural fertilizers is mature compost because it feeds the soil with humus and plant nutrients. Natural fertilizer comes from animal wastes and plants; for example, cow dung, sheep, goat or chicken droppings, urine, decomposed weeds and other plant or animal remains, e.g. waste from preparing food. Good quality compost can be made from organic household wastes in urban areas and be used to grow healthy vegetables in gardens at home or by school environment club or youth group members.

Importance of Compost

Compost is important because it:

- Contains the main plant nutrients – nitrogen (N), phosphorus (P) and potassium (K), often written as NPK;

- Improves the organic matter in the soil by providing humus;

- Helps the soil hold both water and air for plants; and

- Makes trace elements or micronutrients available to plants.

Uses of Compost

Because compost is made up of humus, it can be used for improving soil as follows:

1. It provides plant nutrients that are released throughout the growing season.

   The plant nutrients are released when organic matter decomposes and is changed into humus.

   The plant nutrients dissolve in the water in the soil and are taken in by the roots of the crops.

2. It improves soil structure so that plant roots can easily reach down into the soil.

   In sandy soil the humus makes the sand particles stick together. This reduces the size of the spaces (pores) so that water stays longer in the soil.

   In clay soils, the humus surrounds the clay particles making more spaces (pores) in the soil so the root systems of plants can reach the water and nutrients that they need, and air can also move through the soil.

   Therefore, because heavy clay soils become lighter and sandy soils become heavier, soil that has had compost added to it is easier to work, i.e. to plough and dig.

3. It improves the moisture-holding capacity of soil.

   The humus is a dark brown or black soft spongy or jelly-like substance that holds water and plant nutrients. One kilogram of humus can hold up to six litres of water.

   In dry times, soil with good humus in it can hold water longer than soil with little humus.

In Ethiopia, crops grown on soil with compost can go on growing for two weeks longer after the rains have stopped than crops grown on soil given chemical fertilizer.

When it rains, water easily gets into the soil instead of running off over the surface.

Water gets into the subsoil and down to the water table, runoff and thus flooding is reduced, and springs do not dry up in the dry season.

4. It helps to control weeds, pests and diseases.

When weeds are used to make compost, the high temperature of the compost-making process kills many, but not all, of the weed seeds. Even the noxious weed, Parthenium, has most of its seeds killed when it is made into compost.

Fertile soil produces strong plants able to resist pests and diseases.

When crop residues are used to make compost, many pests and diseases cannot survive to infect the next season's crops.

5. It helps the soil resist erosion by wind and water. This is because:

Water can enter the soil better and this can stop showers building up into a flood. This also reduces splash and sheet erosion.

Soil held together with humus cannot be blown away so easily by wind.

6. Compost helps farmers improve the productivity of their land and their income.

It is made without having to pay cash or borrow money, i.e. farmers do not have to take credit and get into debt like they do for taking chemical fertilizer.

But, to make and use compost properly farmers, either individually or working in groups, have to work hard.

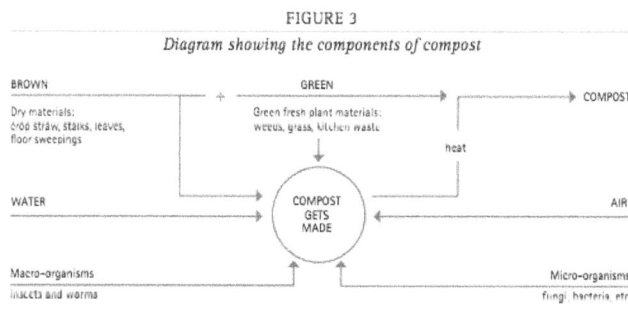

FIGURE 3

Diagram showing the components of compost

Material Required to Make Compost

Plant Materials, Both Dry and Green

1. Weeds, grasses and any other plant materials cut from inside and around fields, in clearing paths, in weeding, etc.

2. Wastes from cleaning grain, cooking and cleaning the house and compound, making food and different drinks, particularly coffee, tea, home-made beer, etc.

3. Crop residues: stems, leaves, straw and chaff1 of all field crops – both big and small – cereals, pulses, oil crops, horticultural crops and spices, from threshing grounds and from fields after harvesting.

4. Garden wastes – old leaves, dead flowers, hedge trimmings, grass cuttings, etc.

5. Dry grass, hay and straw left over from feeding and bedding animals. Animal bedding is very useful because it has been mixed with the urine and droppings of the animals.

6. Dropped leaves and stems from almost any tree and bush except plants which have tough leaves, or leaves and stems with a strong smell or liquid when crushed, like Eucalyptus, Australian Acacia, Euphorbia, etc. However, we have found farmers making good quality compost including stems of Euphorbia.

7. Stems of cactus, such as prickly pear, can be used if they are crushed or chopped up. They are also a good source of moisture for making compost in dry areas. When the compost is made correctly, the spines are destroyed.

Water

Enough water is needed to wet all the materials and keep them moist, but the materials should not be made too wet so that they lack air and thus rot and smell bad. Both too little and too much water prevent good compost being made.

Water does NOT need to be clean like drinking water.

It can come from:

- Collected rainwater;

- Collected wastewater, e.g. from washing pots and pans, clothes, floors, etc.;

- Animal urine; or

- Human urine.

Water can also be collected from ponds, dams, streams and rivers, particularly if men are willing to do it. It is not fair to expect women to collect all the water needed to make compost.

Animal Materials

1. Dung and droppings from all types of domestic animals, including from horses, mules, donkeys and chicken, from night pens and shelters, or collected from fields.

2. Chicken droppings are important to include because they are rich in nitrogen.

3. Urine from cattle and people:

- Catch urine in a container from animals when they wake up and start moving around in the morning.

- Provide a container: like an old clay pot or plastic jerry can – in the toilet or latrine where people can pass or put their urine.

- Night soil (human faeces): almost all human parasites and other disease organisms in human faeces are killed by the high temperatures when good compost is made.

Compost Making Aids – "farmers' Friends"

Micro-organisms (fungi and bacteria) and smaller animals (many types of worms, including earthworms, nematodes, beetles and other insects) turn waste materials into mature compost. These are found naturally in good fertile soils like those from forests, old animal dung and old compost. Adding any of these to new compost helps in the decomposition process.

Adding compost making aids is like adding yeast to the dough to make bread.

The farmers in Ethiopia call these materials the 'spices' to make good compost.

Air

Including dry materials in the compost, e.g. old leaves and stalks, provides space for air to circulate inside the compost. Air is needed because the soil organisms need oxygen.

Heat

Decomposition of organic wastes produces heat. Compost needs to be kept hot and moist so the plant and animal materials can be broken down quickly and thoroughly.

Heat Destroys Most of the Weed Seeds, Fungal Diseases, Pests and Parasites.

The Contributions of the Different Compost-making Materials

A Good Balance of Carbon and Nitrogen

Both carbon and nitrogen are needed to make good compost. They are used by the micro-organisms to grow and multiply, and to get energy. Some of the carbon is converted to carbon dioxide, and this escapes to the atmosphere. Most of it remains and becomes humus, and the nitrogen becomes nitrates. Methane is not produced if there is a good supply of air to the organisms carrying out the decomposition process.

Materials with good nitrogen content help in making good compost, but they should be less than the carbon-containing materials. Carbon-containing materials should always be more than those containing high nitrogen. A good balance of carbon and nitrogen is needed to make good compost. Table below gives the carbon-to-nitrogen balance for some types of composting materials.

Tables

The nitrogen and carbon content of some selected composting material.s

Type of composting material Nitrogen content (%) Carbon-to-Nitrogen ratio (C:1N)

Type of composting material	Nitrogen content (%)	Carbon-to-Nitrogen ratio (C:1N)
Urine	15–18	0.8:1
Blood	10–14	3:1
Horn	12	not found
Bone	3	8:1
Chicken manure	3–6	10–12:1
Sheep manure	3.8	not found
Horse and donkey manure	3.8	25:1
Manure in general	1.7	18:1
Manure from animal pens = farmyard manure (FYM)	2.15	14:1
Maize stalks and leaves	0.7–0.8	55–70:1
Wheat straw and chaff	0.4–0.6	80–100:1
Fallen leaves	0.4	45:1
Young grass hay	4	12:1
Grass clippings	2.4	20:1
Straw from peas and beans	1.5	not found

With Nitrogen as 1, high figures for the carbon in the carbon-to-nitrogen column indicate high carbon content. These items are good for making compost.

Items with low carbon content, like urine and chicken manure, are useful to provide nitrogen. But they must be mixed with materials with high carbon content.

1. When there is enough air and moisture in the compost, nitrogen-containing materials are broken down and the nitrogen is changed to nitrates that can be used by plants.

2. When there is too much water and little air, the nitrogen is changed into ammonia.

    This is a gas that escapes from the compost, and gives the compost a bad smell.

3. When there is a bad smell, the compost needs to be turned over bringing the top to the bottom and the bottom to the top, and mixing in more dry materials and some good soil. This puts more air into the compost, which stops the process of making ammonia so that proper mature compost can be made.

The Contributions of Dry and Green Plant Materials

Dry materials give structure to the compost making process; they provide space for air to circulate so that the micro-organisms can be active and make heat.

Green plant materials provide moisture for compost making; they give water and nutrients to the micro-organisms so that they multiply and break down the organic materials into humus.

The Importance of Good Water/Moisture and Air Balance

Water is essential for compost preparation.

1. Sufficient moisture helps for quicker decomposition because it is essential for micro-organisms to be active.

2. Excess water causes rotting of the materials and creates a bad smell.

3. Without enough moisture the decomposition process slows down and the materials will not be changed into compost.

This shows that moisture and air must be balanced to make good compost. Farmers quickly learn how to judge the amount of water needed to be added in making compost.

The Importance of Air

Compost should have sufficient air.

1. When there is sufficient air, oxygen enters the compost heap. When there is enough oxygen, special bacteria can convert nitrogen into nitrate, the materials are decomposed properly and there is a good smell.

2. If there is not enough air and too much water, the nitrogen is converted into ammonia. The ammonia escapes into the air removing nitrogen from the compost and making it smell bad.

3. If there is excess air and too little water, the materials dry up and do not decompose to become compost.

Quality Compost with Animal Dung and Urine

1. Animal dung contains water, nitrogen, phosphorous and potassium, as well as micro-nutrients.

2. Animal dung and urine are very necessary to prepare good quality compost – urine especially is high in potassium and nitrogen.

3. Both dung and urine help to produce a high temperature so that the materials decompose into compost easily.

4. Urine, in particular, accelerates decomposition.

References

- Y.P. Chen; P.D. Rekha; A.B. Arun; F.T. Shen; W.-A. Lai; C.C. Young (2006). "Phosphate solubilizing bacteria from subtropical soil and their tricalcium phosphate solubilizing abilities". Applied Soil Ecology. 34 (1): 33–41. doi:10.1016/j.apsoil.2005.12.002.

- What-biofertilizer-dr-tohid-nooralvandi: linkedin.com, Retrieved 12 March 2018

- Advantages-and-disadvantages-of-biofertilizers-13404698: hunker.com, Retrieved 16 April 2018

- Park, J. H., Bolan, N., Megharaj, M., & Naidu, R. (2011). Isolation of phosphate solubilizing bacteria and their potential for lead immobilization in soil. Journal of hazardous materials, 185(2), 829-836.

- Production-of-various-bio-fertilizers-microbiology-66873: biologydiscussion.com, Retrieved 27 May 2018

- Nitrogen-Biofertilizers-and-New-Microbial-Strains-360: biotecharticles.com, Retrieved 28 July 2018

Permissions

Index

www.ingramcontent.com/pod-product-compliance
Lightning Source LLC
Chambersburg PA
CBHW082010190326
41458CB00010B/3145